Implementing Virtual Design and Construction using BIM

Implementing Virtual Design and Construction using BIM outlines the team structure, software, and production ecosystem needed for an effective Virtual Design and Construction (VDC) process through current real-world case studies of projects both in development and under construction. It provides the reader with a better understanding of the successful implementation of VDC and Building Information Modeling (BIM), and the benefits to the project team throughout the design and construction process. For readers already familiar with VDC, the book will provide invaluable examples of best practices and real-world solutions.

Richly illustrated in color with actual VDC documentation, visualizations, and statistics, the reader is shown the real processes undertaken and outputs generated when working on high-profile building information models. Online animations, interviews with practitioners, and downloadable templates, forms, and files make this an interactive and highly engaging way to learn a crucial set of skills.

While keeping up with current industry practice is a minimum requirement, this book goes further by helping practitioners prepare for the next level of virtual design and construction. This is essential reading for project managers, construction managers, architects, design managers, and anybody with a role in BIM or virtual construction.

Lennart Andersson is the director of Virtual Design and Construction (VDC) at LiRo. He studied engineering in Sweden and received a master's degree in architecture in the USA. He is a licensed architect with sixteen years of experience applying virtual design and construction methodologies on a wide variety of projects. Lennart is spearheading the firm-wide adoption of VDC at LiRo. Examples of projects he has been involved in are East Side Access, City Point Brooklyn, New York Public Library, and a number of other large-scale projects in New York City. He is a visiting professor at Pratt Institute in New York, where he teaches an advanced, collaborative VDC studio between architects and construction managers in partnership with the New York City Department of Design and Construction. He has hosted several seminars covering advanced VDC use in design, construction, and operations.

Kyla Farrell worked as a VDC project manager at the LiRo Group from 2012 to 2014. Prior to LiRo, she was employed by Gehry Technologies, Zaha Hadid Architects and SHoP Construction, where she developed and managed parametric 3-D models for the design, fabrication, and coordination of complex projects. Kyla received her first degree from Stanford University and holds a master's degree in architecture from the Southern California Institute of Architecture. She currently works as a software engineer in the San Francisco Bay Area.

Oleg Moshkovich is a VDC product architect at the LiRo Group. He is responsible for the design and implementation of virtual construction processes on projects of increased complexity. His main engagement at LiRo is in the design and implementation of VDC on the East Side Access project. His multidisciplinary and multicultural experience allows him to envision and to communicate technology-driven, integrated project solutions. His experience includes engineering, architecture, web development, and project controls. He holds a master's degree in engineering in product architecture from Stevens Institute of Technology, USA.

Cheryle Cranbourne is a VDC specialist at the LiRo Group, involved in modeling complex and large-scale projects. She is also responsible for creating animations and informational graphics for project development and marketing purposes, such as the experience simulations of the East Side Access and City Point projects. She has a diverse and multicultural design background, having worked in interior architecture, advertising, and graphic design roles in the USA and Asia. Cheryle received a B.Sc. in communications from Boston University, and a master's degree in interior architecture from the Rhode Island School of Design, USA.

Implementing Virtual Design and Construction using BIM

Current and future practices

Lennart Andersson, Kyla Farrell,
Oleg Moshkovich, and
Cheryle Cranbourne

Routledge
Taylor & Francis Group

LONDON AND NEW YORK

First published 2016
by Routledge
2 Park Square, Milton Park, Abingdon, Oxon OX14 4RN

and by Routledge
711 Third Avenue, New York, NY 10017

Routledge is an imprint of the Taylor & Francis Group, an informa business

British Library Cataloguing-in-Publication Data
A catalogue record for this book is available from the British Library

Library of Congress Cataloging in Publication Data
A catalog record for this book has been requested

ISBN: 978-1-138-01994-2 (hbk)
ISBN: 978-1-315-65707-3 (ebk)

Typeset in Myriad Pro
by Florence Production Ltd, Stoodleigh, Devon, UK

Contents

Illustrations

Preface

Information technology is rapidly connecting all corners of the world, dissolving physical barriers and enabling previously impossible global interactions. As per Moore's Law,[1] computer hardware is evolving at a breathtaking pace; a mobile phone today is in some ways more powerful than a building-sized supercomputer was some 30 years ago. Furthermore, cloud computing has opened up new forms of communication and channels for sharing information around the globe. Cloud computing provides an abundance of available processing power, which has an enormous impact on processes and workflows across all areas of human endeavor.

We are entering the era of the Third Industrial Revolution. Technological innovations in manufacturing such as robotics are radically changing how products are made. Manufacturing techniques such as 3-D printing make it possible to create products locally and on demand, rather than stockpiling or shipping them across the world. Social networks and the interconnectedness of information technology are fostering new ways of thinking and working collaboratively. Crowdsourcing is making it possible for groups of physically disparate people to team up and work on projects and collectively arrive at solutions. It is more beneficial for individuals to share rather than "own" information, as the expiration date for some knowledge is drastically shorter now than in the past. Continuous learning must be incorporated into everyday workflows to stay on track with current progress.

GLOBAL PERSPECTIVE

As we enter an age in which many more people can produce drastically more product than was previously possible, we face enormous challenges in tending to the limited resources of this planet. Around the world, we are already depleting the resources we depend on.

The Industrial Revolution grew out of a view that we have infinite resources at our disposal, so the process of design focused solely on manufacture and usage. Very little thought was given to the impact of production on the environment, or what happens to something once it has served its purpose. Today, a different way of thinking must prevail; we need to put what we do into context and realize the interconnectedness of the systems in which we operate.

BUILDING INDUSTRY

The global building industry is one of the largest industries in the world and will grow from approximately an $8 trillion industry in 2013 to a $12 trillion industry by 2020.[2] Yet, building construction is still quite often a low-tech environment that can be extremely inefficient and wasteful. Indeed, it may be the only industry that has actually declined in efficiency over the past 20 years.

As building requires an enormous amount of resources, the industry has substantial effects on the environment. This book addresses the rapidly evolving technological tools that will make it possible to understand and change how things are built, and to streamline construction processes and minimize waste. The tools themselves will not solve our environmental quandary, but they enable us to visualize and solve complex problems. These tools bring transparency to the industry and expose the myriad interconnected issues involved in the building process. The industry can dramatically reduce its amount of waste by efficiently utilizing virtual building technologies.

BUILDING VIRTUAL

This book illustrates how technology can be successfully applied on a range of projects from an academic and theoretic perspective, and demonstrates how Virtual Design and Construction (VDC) actually interfaces and functions in the real world through case studies in New York City. The case studies included here are supported by reference chapters, which describe the tools and settings that help ensure the success of a VDC project.

As we are constantly developing VDC tools and workflows, please refer to our website for the most up-to-date examples and files at www.buildingvirtual.net.

ACKNOWLEDGMENTS

We would like to thank everyone who helped us on the projects described in this book, especially LiRo's CEO Luis Tormenta for believing in our department from the very beginning. Additionally, our colleagues Michael Bailey, Michael Burton, Vikas Wagh, and Vincent Valdemira who provided us with valuable industry knowledge, time, and resources. Our clients also deserve a big thanks, including Mark DeBernardo at MTA Capital Construction, for being our advocate when we started at East Side Access. Many thanks go out to Arta Yazdanseta and Harriet Markis for giving valuable feedback on early book concepts. An additional thanks to Brian

Guerin and Matthew Turpie at Routledge; our editor, Naveen Kumar; and our friends and family who read drafts and put up with our absence as we put this book together: Suzanne Bennett, Matt O'Neill, Vijay Pandurangan, Tan Shin Bin, and Andrea Lausevic.

Last but not least, we would like to thank everyone on the LiRo VDC team integral to the success of these amazing projects: Fernando Vazquez, Scott Blond, David Deiss, Izzet Keskintas, Aditi Patel, Jorge Berdecia, David Wu, and Brian Szeto.

NOTES

1   Moore's Law is named after Intel co-founder Gordon Moore, who predicted in
    1975 that every two years the number of transistors in chip elements will double.
    So far, this prediction has proved to be true. Stephen Shankland, "Moore's Law:
    The Rule That Really Matters in Tech." CNET, October 15, 2012. Web. December 23,
    2014.
2   David R. Schilling, "Global Construction Expected to Increase by $4.8 Trillion by
    2020." Industry Tap. March 08, 2013. Web. November 10, 2014.

1 Introduction

Disruptive new technologies are transforming all facets of the built environment. Virtual Design and Construction (VDC) is the implementation of these technologies and processes. Understanding this emerging field is essential for all professionals working in Architecture, Engineering, and Construction (AEC).

The success of VDC depends not only on technology, but on the skills and knowledge of people who initiate, design, construct, and operate projects using a wide variety of professional tools. The ambition of this book is to communicate how powerful new tools significantly improve the process of building, as well as the quality of resulting buildings.

The AEC industry refers to much of what we discuss here simply as Building Information Modeling, or BIM. We find "BIM" to be an inadequate description of the workflows we are developing as VDC professionals. For the purposes of this book, the result of BIM or "BIM model" will be referred to simply as an "information model." For activities incorporating use of an information model, we use the terms VDC process (or methodology), VDC service, or VDC product. VDC processes are workflows that incorporate the information model and integrate previously disconnected aspects of design and construction. VDC processes seek to apply new technologies to the AEC industry and link all the work being done by the project team into the information model. The information model acts as a hub. VDC services are specific services unique to VDC, such as clash detection, 3-D scanning, tracking, or information model authoring. A VDC product is the deliverable resulting from a VDC service, such as a point cloud, a systems coordination model, database, or a constructability logistics animation.

While the concept of BIM has its roots in the early beginnings of computer technology, it was not until the personal computer became powerful enough to drive the data and graphics in real time that 3-D models became a useful tool. An information model simulates the geometry and data of an environment, unlike Computer Aided Drafting (CAD), which is merely a representation, like a drawing on paper. The information model is a virtual, geometrical, spatial relational database. It keeps track of data as it relates to specific geometry and location. Many types of data can be linked to a virtual object, and there are many possible ways to use and analyze the data contained in the model.

An information model is powerful because it allows all of the data surrounding a building project to be centralized into one ecosystem that all participants can share. This centralization mitigates problems associated with the fragmentation of data inherent in the traditional design and construction process. For example, someone viewing color-coded 3-D models instead of black-and-white line drawings gains a much better understanding of the project at hand, as relationships between different components are more clearly visible. Using information models thus minimizes the risk of misunderstandings and subsequent conflicts.

The case studies in this book are written from the perspective of our experience working in the VDC department within the LiRo Group, a Construction Management, Architecture, and Engineering firm headquartered in Syosset, NY. LiRo is a professional, full-service design and construction management firm ranked among the nation's Top 20 CM firms by Engineering News Record in 2014. The VDC department operates out of its own office in Manhattan. In addition to the VDC group, LiRo's current workforce of over 650 personnel includes licensed professional engineers, architects and field staff experienced in design, pre-construction, construction inspection and supervision, CPM scheduling techniques and computerized logging and document control systems. The staff also includes experienced value engineers, certified cost estimators, and LEED accredited professionals. The construction management team enlists in-house environmental, structural, traffic, and civil engineers, hazardous material specialists, PLA consultants and database developers, among others, to respond to any technical need that may arise on a project. From our vantage point in the VDC department, working with LiRo's full spectrum of designers, engineers, and constructors, we have a deep understanding of how the various processes of a building project relate to each other.

The construction manager's (CM) main role is to ensure that the intended design is built in the best possible way, at the lowest cost and in the most time-efficient manner. The tools a CM uses apply mostly to means and methods, such as planning and tracking the construction of the project. A CM ensures that all parties understand their scope and responsibilities through contractual documents. Specific services rendered include specification authoring, sequencing and scheduling, cost estimating, constructability review of the intended design, creating staging plans, tracking and reporting progress, enforcing site safety, quality assurance and control as well as cost-related tasks, such as value engineering and administration. All these services can be greatly improved with VDC processes. The CM might actually be one of the greatest beneficiaries of VDC, as the transparency it affords helps the CM understand and monitor every aspect of the project.

VDC will only continue to expand as a discipline, becoming a further integrated part of the AEC process.[1] New technologies and innovations are constantly being devised to address the many inefficiencies in current professional practice. As VDC professionals, we are interested in the rapid advances being made in the development of new technologies that facilitate a bidirectional link between the real and the virtual, providing a platform for better decision making. 3-D scanning, 3-D printing, sensors, prefabrication, automation, and robotics are among the many exciting innovations being developed. At its core, VDC ultimately seeks to bridge the expertise gaps between design, construction, and operations; to realize facilities that are dramatically less wasteful both in assembly and usage; and to create buildings that function to serve their occupants throughout the complete usage lifecycle.

NOTE

1   Forty percent of US owners and 38 percent of UK owners expect that more than 75 percent of their projects will involve BIM in just two years. McGraw Hill Construction, Marketing Communications, "U.S. and U.K. Building Owners Expect to Increase Their Involvement with BIM in the Next Two Years," Market Watch. October 13, 2014. Web, October 24, 2014.

2 The Practice of VDC

VDC is an interdisciplinary practice in which data is centralized, typically within a 3-D information model, allowing for increased efficiencies and deeper project understanding and analysis. VDC is a shift from mere representation of project information as in a 2-D design process to detailed simulation, from a linear design and construction process to a concurrent process with live feedback loops. Implementing a functional VDC practice requires an understanding of the building process, structure and professional culture both at the project and enterprise level.

VDC processes are workflows that incorporate the information model and integrate previously disconnected aspects of design and construction. VDC processes seek to apply new technologies to the AEC industry and link the work done by the project team to the information model. The information model acts as a central hub in the VDC workflow. VDC services are specific services unique to VDC, such as clash detection, 3-D scanning, tracking or information model authoring. A VDC product is the deliverable resulting from a VDC service, such as a point cloud, a systems coordination model, database or a constructability logistics animation.

VDC services can be utilized throughout the entire design and construction process. If VDC services simply run parallel to traditional workflows, they don't provide the optimal benefit to a project. VDC services must be integrated into the traditional trades and everyday workflows to be effective. Every member of the team needs a certain level of understanding regarding VDC in order to innovate and improve existing practices. Successful VDC implementation requires a thorough understanding of how things are done in theory as well as practice. Understanding the team's existing structure of decision making is crucial to implement effective new practices.

A VDC department's success depends not only on the talent of its team and strong process awareness, but also on clear organization. The structure of the VDC practice should evolve with each project, simplifying initial deployment, and incorporating lessons learned from previous projects, which are captured as a set of pre-formatted templates, databases, and a clearly organized file tree. Clear naming conventions and correctly implemented interoperability standards are the conduits that connect VDC to traditional AEC workflows and are addressed further in Chapter 5 ("Reference Documents").

2.1 VDC Services

VDC services broadly fall into three categories: implementation, production, and support services. Implementation includes consulting and educating a project team on the integration of VDC into a project's design-through-construction workflow. Writing up VDC Specifications and Implementation Plans, and maintaining the overall quality of models all fall into this category. Production is the work of creating deliverables and output from the various types of specialized information models outlined in section 2.2, each of which supports specific VDC services. Support services are those that include using the model to solve specific project issues that emerge throughout the course of the existing design and construction workflow. Examples of such services include litigation support and risk workshops.

IMPLEMENTATION SERVICES

VDC Specifications

Specifications provide the rulebook for a project. They set expectations and outline how work should be performed. A good VDC Specification states what the information model should include and to what level of detail, as well as major information model deliverables for all phases of a project, from early design to the facility's final operations. The Specification should also reference the related global standard for level of model development. Global standards are in development for BIM and VDC. In the USA, the most prominent of these are the National BIM standards and the Level of Development document.[1] (For these standards, readers are referred to Chapter 5, "Reference Documents," which includes an example VDC Specification document.) Contracting bids are submitted based on the provided specifications.

How VDC should be integrated into a project greatly depends on how a team is organized. Every project has different requirements, and team organization varies based on the project's typology, size, complexity, client, location, phasing, and other requirements. A team's level of sophistication is an additional factor to consider. For example, where some parties are not sufficiently capable of operating information modeling software that might negatively affect how, or even if, VDC is incorporated into the process.

Providing detailed specifications that outline the implementation of a VDC process is extremely important. Any omission will likely adversely affect other aspects of the project; successful collaboration requires clearly defined standards. Essential standards for an information model include naming conventions, file structure, software workflows, component definitions, model completeness, and data output. Standardized formats for sharing 3-D data ensure the consistency and compatibility of both internal and external sharing.

Procedures for the implementation of new technology are essential, including the choice of software platform(s), identification of individuals who require software training, and analysis of existing technological infrastructure, including upgrades to targeted computers.

VDC Implementation Planning

The VDC Specification should require the team member responsible for model authoring and model coordination to produce a VDC Implementation Plan. The plan is typically the responsibility of the design team during the design phase, the contractor during the construction phase, or the VDC consultant who is supervising coordination in either phase. A clearly defined Implementation Plan that answers all requirements in the specifications is essential for successfully integrating VDC. It not only outlines in detail how the VDC processes will be implemented and what software and hardware will be used, but also proves if the contractor has VDC capabilities. The Implementation Plan should include the contractor's project team structure as well as 3-D model standards with naming conventions, file organization, software workflows, component definitions, model completeness, and data output definitions. The Implementation Plan needs to be reviewed and approved by the CM, the design team and the owner's representative prior to commencing work. A sample Implementation Plan is included in Chapter 5.

VDC Training

In order to ensure that the VDC workflows function well, all involved parties need to have sufficient knowledge in using VDC. At the inception of a project the skill set and required training need to be assessed. It is effective to have the party that is managing the VDC process to organize the training rather than outsourcing it.

PRODUCTION SERVICES

Visualizations

Visualizing the model is sometimes a separate service, but often it is an integrated part of other VDC services, such as 4-D scheduling. Traditional visualization of project design through perspective drawings is almost as

Figure 2.1.1 (overleaf)
Diagram of VDC models and services throughout typical AEC project phases

PHASE

CONCEPTUALIZATION

DESIGN

PROCUREMENT

PRE-CONSTRUCTION

INFORMATION MODEL

Design Intent Model

Construction Management + VDC Models

Construction Models

CM + VDC MODEL TYPE

3-D Information Model	Visualization Models	Coordination Models
Model Quality Control	Experience Simulation	Design Coordination
Constructability Studies	Contract Scope Visualization	CM Coordination
Virtual Mockups	3-D Printing	Construction Coordination
Field Capturing	Systems Visualization	Interference (Clash) Detection

VDC SERVICES

VDC Specifications

VDC Implementation Plans

VDC Training

Worker Training & Safety

Construction Quality Assurance

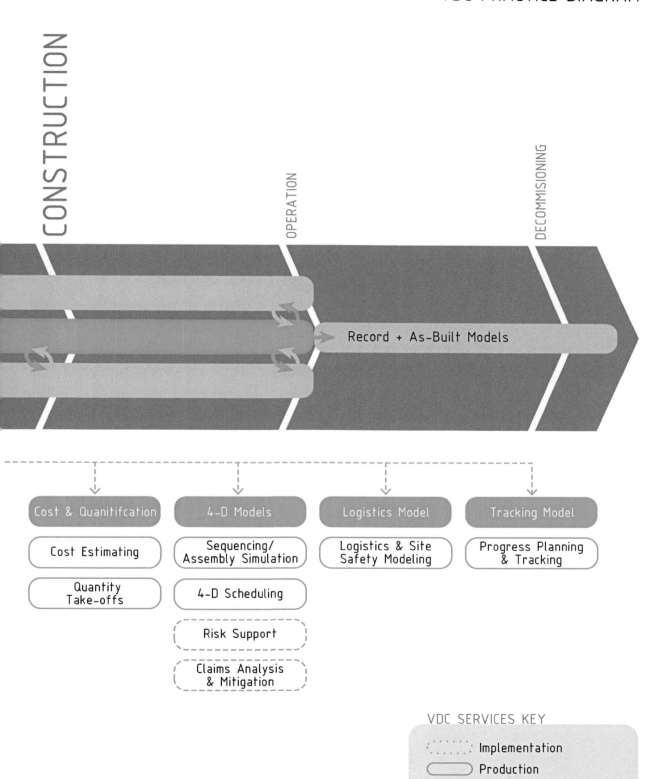

VDC PRACTICE DIAGRAM

CONSTRUCTION

OPERATION

DECOMMISIONING

Record + As-Built Models

Cost & Quanitifcation

Cost Estimating

Quantity
Take-offs

4-D Models

Sequencing/
Assembly Simulation

4-D Scheduling

Risk Support

Claims Analysis
& Mitigation

Logistics Model

Logistics & Site
Safety Modeling

Tracking Model

Progress Planning
& Tracking

VDC SERVICES KEY

Implementation
Production
Support

Figure 2.1.2
A still frame from a walk-through animation

old as the art of building itself. VDC tools make it possible not only to visualize specific views, but also generate performance-driven visualization and the construction of the building in sequence. Images, animations, live virtual walk-throughs, and experience simulations are gradually becoming standard project requirements. Visualization helps all parties quickly understand the project, especially participants who are not skilled in reading 2-D documentation and specifications. Identifying issues becomes crystal clear, and all parties can take part in the discussion to resolve them, rather than spend time and energy establishing the topic of discussion.

Depending on the need, it is possible to generate various types of visualization results, ranging from highly realistic experience animations that convey the final experience of the facility, to color-coded coordination studies meant to communicate concepts, discrepancies, performance criteria, issues, and interferences. Each type of visualization should be tailored to its specific audience.

Experience Simulation

A walk-through of the final constructed design should be shown with the highest possible realism to convey as closely as possible the real experience of place.

Systems Visualization

Visualizations used for systems coordination should be color-coded by subsystems, so some elements can be made transparent or turned off to bring relevant information to the forefront. A sophisticated VDC team tailors each visualization experience to include exactly the necessary data and highlight what is of interest. For some types of visualization it is best to provide both 2-D and 3-D representations, as 2-D helps with location information such as room, grid, zone, and levels.

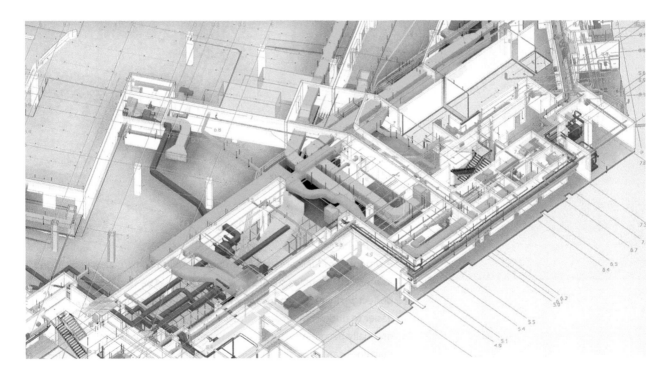

Figure 2.1.3
An example of systems visualization

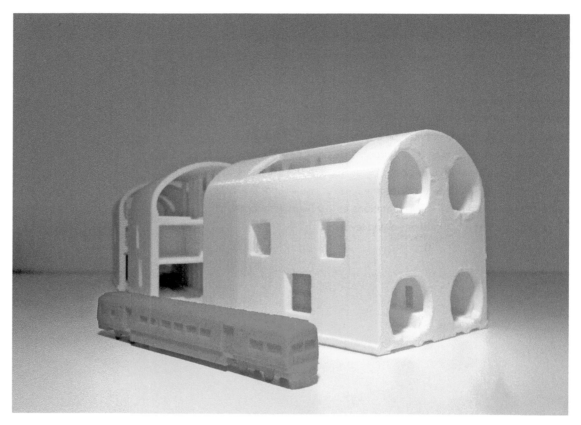

Figure 2.1.4
A 3-D printed prototype produced for an infrastructure project

3-D Prints

3-D printed models are useful for understanding construction sequencing as mockups; complex shapes or connections can be studied in detail. For project participants who are not fluent with computers, a 3-D print is easier to understand than a virtual model. 3-D printing technology is gradually becoming a tool for real manufacturing, such that large machines can produce real components of a building directly from the information model.

Contract Scope Visualization

3-D modeling is a great platform for the contractor to quickly understand the scope of a project, particularly for projects that involve many contractors and subcontractors, with potentially complex interfaces and possible overlaps between different contract scopes. An information model can be used to clearly and precisely visualize the scope of different contractors' responsibilities, tie scope and costs to associated quantities extracted from a model, and highlight critical interfaces between contracts. This capability removes ambiguity and potential inconsistencies regarding different parties' responsibilities, and helps contractors to provide bids that correspond correctly to the proposed work.

Proposal Presentations

Developers often issue Requests for Proposals (RFP) in order to select winning contracts for Design, Constructions Managers, and Contractor. Proposals include documents that illustrate their understanding of the project as well as proposed method of execution. This can include design, staging, phasing, access requirement, scheduling, team configuration, and a range of other qualifications. To help convey these concepts, VDC products can be included as an integrated part of any job proposal. This helps not only in communicating how the job would be run, but also demonstrates how VDC would be used throughout the project as an integrated part of everyday workflows.

Field Capturing

With VDC, capturing existing spaces is usually done by laser scanning, creating an extremely accurate 3-D point cloud that includes every inch of visible surface area. This cloud can be linked to the information modeling authoring software and used to model the space or validate existing information models. Laser scanning is quickly becoming the preferred method of recording existing information, particularly as the equipment becomes faster and more cost-effective. Newer scanners use accurate colors in order to visualize a realistic environment. Information model authoring software is evolving such that architectural, structural, and systems components can now be modeled almost automatically, directly from the point cloud with the use of specialized software plugins.

Point clouds generated from laser scanning consist of millions of points, which results in gigantic amounts of data. A complete set of point clouds can exceed terabytes (TB) of data, so currently there is no practical way of distributing them over the internet. Often scanning consultants prefer to deliver point clouds using Blu-ray disks, as they hold up to 128 gigabytes (GB) of data that cannot be altered. It is important to keep in mind that saving files of that size can cause major issues with IT infrastructure because they take up so much drive space. Backup procedures might not be set up to handle such enormous amounts of data. Therefore projects that include modeling from a point cloud are often handled as separate tasks, with specific drives set up solely to handle point cloud data.

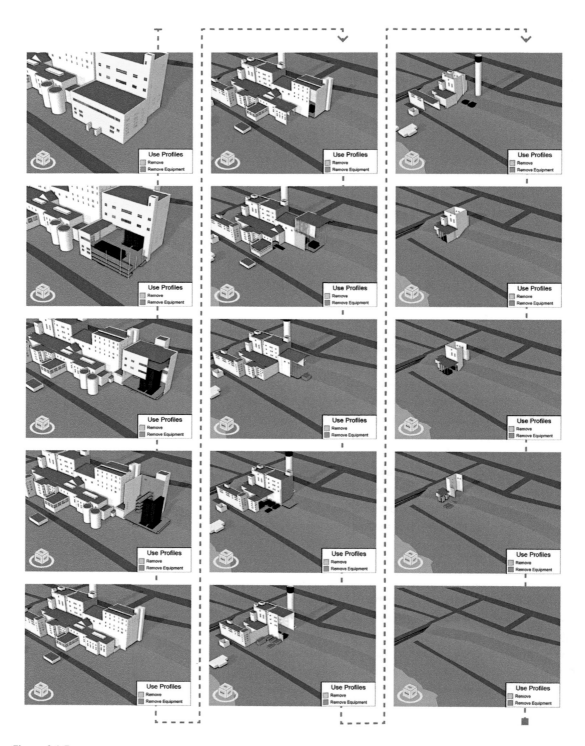

Figure 2.1.5
Proposal graphics: stills from an animation conveying the demolition of a power plant

Figure 2.1.6
A series of images exported from a point cloud of the Biltmore Room in Grand Central Station, New York City

X-ray methods can be used to detect objects hidden underground or inside a wall, but they are rather costly and only detect certain materials and reach certain depths. However, all the methods mentioned above have one major limitation: They do not show the composition of an element, only its geometric representation in space. For example, a 3-D scan cannot differentiate between a pipe and some other cylindrical object, such as a round column, or between types of pipes.

Constructability Studies

The information model is the perfect tool for constructability reviews. Traditional 2-D documents are representations of the intended construction and do not describe every condition; they merely represent "slices" of the project. Any condition beyond those cuts is not described. As the information model includes every single location, and all the trades can be linked into one file, relationships in the model can be studied in greater detail. Virtually constructing the building helps identify a vast range of issues, especially when the project can be visualized over time, linked to a schedule, and its quantities used for cost estimating. This quickly reveals issues that can then be rectified before construction begins.

Figure 2.1.7
A Constructability Review report

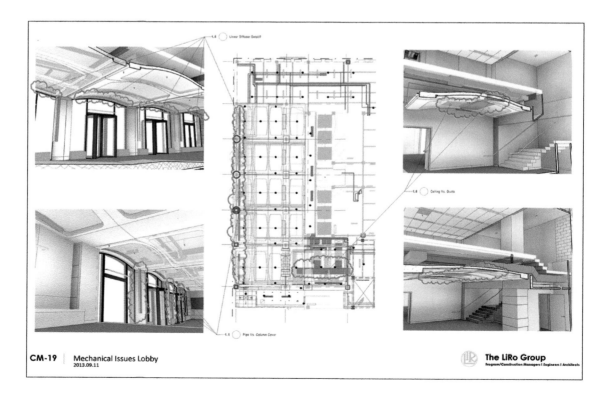

CM-19 | Mechanical Issues Lobby
2013.09.11

The LiRo Group
Program/Construction Managers | Engineers | Architects

Virtual Mockups

Construction and complex building analysis mockups can include detailed modeling of foundation and excavation scenarios, formwork, analysis of construction sequence alternatives, virtual mockups of alternative primary structure, and curtain wall system designs. A much higher level of detail can be modeled to analyze more complex conditions, such as how exactly the different parts of an assembly fit together. Typical areas of interest are the joints and edges between assemblies, which require collaboration between multiple trades. These specialized models can also be 3-D printed to increase understanding of the relationships between pieces. The different parts can be color-coded and made to be easily assembled and taken apart.

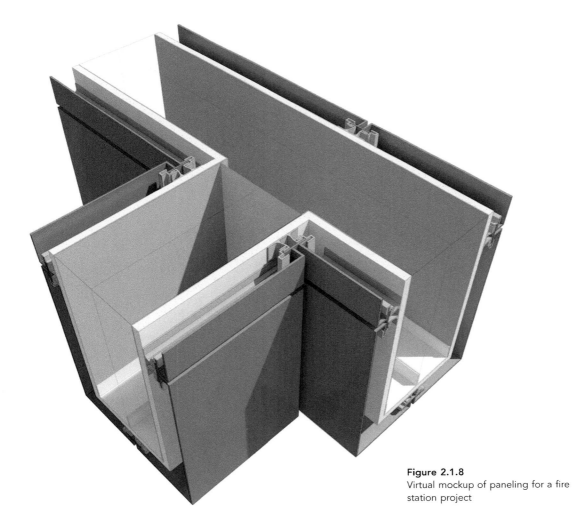

Figure 2.1.8
Virtual mockup of paneling for a fire station project

Figure 2.1.9
This sequence of images
communicates a beam installation
proposal. Two 20-foot beams
maneuvering into place within a
complex existing structure

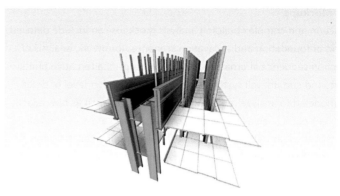

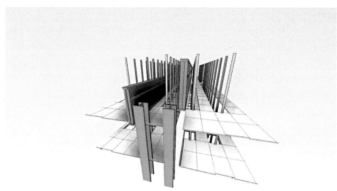

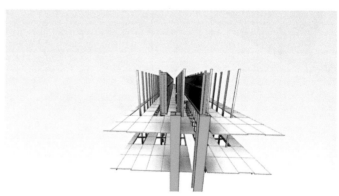

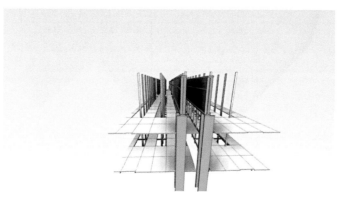

Sequencing and Construction Assembly Simulation

Geometric data in a 3-D model can be augmented with phasing and constructability information to run what-if scenarios for construction sequencing and resource allocation, which can help resolve potential problems ahead of time or in tandem with work being performed on-site. During the construction phase, on-site modelers can assist project staff with detailed analysis, quality control, and validation for daily construction tasks. Examples of this type of work include detailed site modeling of formwork, scaffolding, reinforcing bar and penetration details, as well as sequencing of assembly. For larger elements, especially complex prefabricated systems, it is useful to animate the delivery to ensure there are no spatial conflicts or issues with the sequence of installation.

Information Model Quantification and Cost Estimating

An information model generates accurate quantities that can be linked to a cost database, which then computes a cost estimate. This is fundamentally different from conventional "manual" calculation methods, which mainly consist of counting elements and areas from partial drawings that rarely describe the complete project. For the VDC workflow to function properly, reviewing the completeness, quality, and accuracy of the information model is essential. It is also important for the estimator to know the model's level of development, as some items may not be included. The takeoffs will be instant, but will only be as accurate as the model. A program called Innovaya produces accurate cost estimates by linking to Timberline, a cost database which contains up-to-date cost statistics.

Certain built-in quality control measures help the cost estimator ensure that the model is generating accurate quantities. Takeoff methods mainly rely on the Uniformat system for construction classifications. The components can then utilize different assembly calculations in order to achieve the correct quantity and unit.

Systems Coordination and Interference (Clash) Detection

Systems coordination can happen in two phases: initially during the design phase and then during construction, when subcontractors are coordinated for manufacturing and installation. Clear methods and guidelines for systems coordination are paramount in order to collaborate in the most efficient manner. Systems are modeled in sequence, based on a hierarchy of importance and the flexibility of each system. Every system and subsystem is color-coded for ease of visual understanding. A comprehensive template is used to track a wide range of different interferences. The location of each interference is saved in the model, so

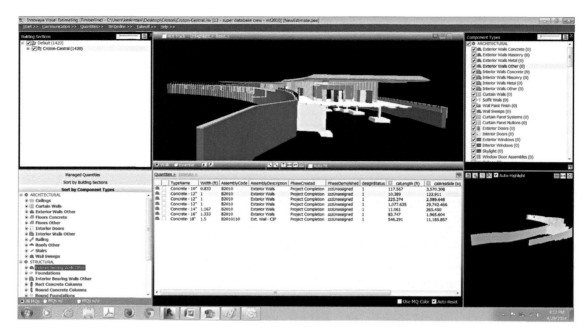

Figure 2.1.10
A screenshot of cost-estimating software Innovaya. The highlighted blue bearing walls are being reviewed for quantities and cost

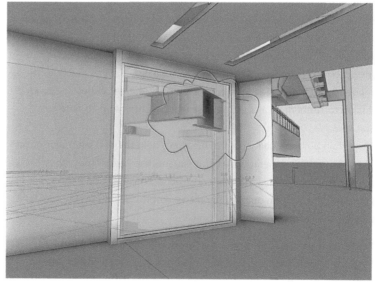

Figure 2.1.11
An example of interference (clash detection)

that it can be actively used in the resolution of each issue. Interferences are also statistically recorded and broken down per trade, so the resolutions can be tracked over time. During the construction phase, the VDC team works closely with systems contractors to ensure the project is coordinated and modeled in the most efficient manner. As most trade modelers use CAD-based systems, each participant's layer standard is reviewed so it interfaces seamlessly with the ongoing tracking of interferences in the Coordination Model.

Figure 2.1.12
Systems subcontractor coordination

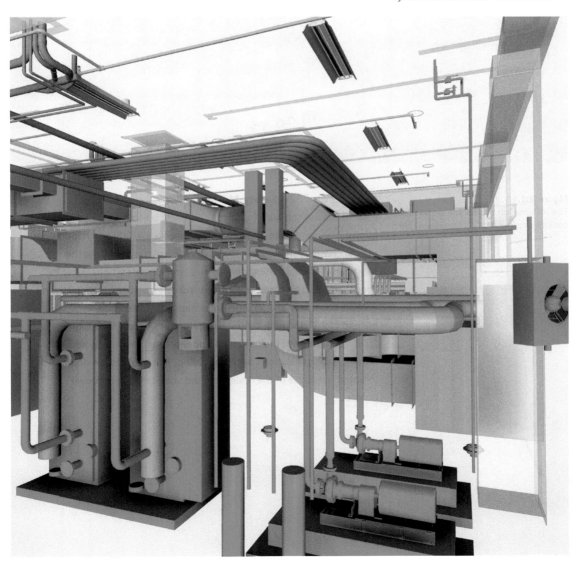

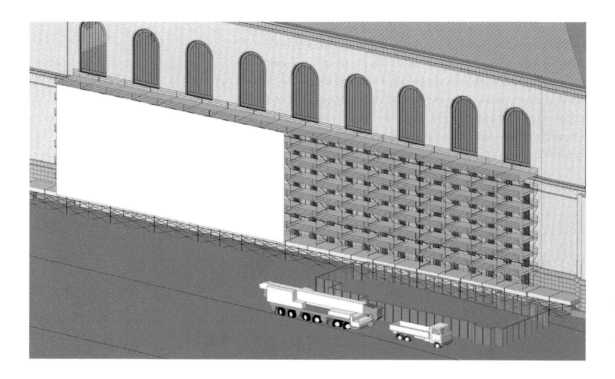

Figure 2.1.13
The information model can be used to
convey site logistics

Logistics and Site Safety Modeling
Code-checking software (such as Solibri) can perform certain tests on
information models to measure code compliance. City and national
planning commissions around the world increasingly accept the
submission of information models to demonstrate compliance with
specific codes. For example, the New York Department of Buildings now
accepts site safety and logistics information models as a submission
instead of 2-D documentation. The department provides a logistics
Revit template with preloaded components such as cranes, access
arrows, staging locations, and scaffolding. This enables the department
to automate some of the reviewing and speed up the approval
process.

4-D Scheduling

4-D scheduling tools have evolved tremendously in recent years. The concept is to link scheduling activities with the information model in order to visualize the sequence of construction. This method is revolutionizing how complex projects are planned, as it gives visual feedback to the scheduler, making relationships between the different assemblies easier to understand.

Objects can be color-coded based on numerous criteria, such as activity type, company, trade, or whether they are late or early. Different versions of the schedule can also be run to compare changes. 4-D scheduling is an effective way of auditing the schedule for missing activities and incorrect sequencing.

For a schedule built without the use of an information model, linking all the geometry to the relating activities can be time-consuming. Once a 4-D schedule is set up, it requires little work to maintain. When updates are made, it's easy to run comparisons with the baseline schedule and spot late or early activities.

A scheduler versed with VDC can actually use an information model to build the schedule, instead of just using the model to visualize the schedule. With a well-organized information model, some activity generation can be automated in order to speed up the process. The leading 4-D scheduling software is Synchro; a fully integrated scheduling software designed for industry professionals, it can read and write Primavera files, a common format of conventional scheduling software. Autodesk Navisworks has the capability to visualize schedules, but its options are limited and it does not provide a seamless editing experience.

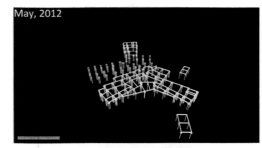

May, 2012

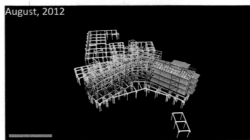

August, 2012

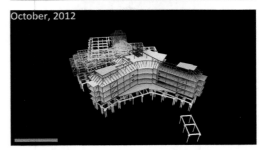

October, 2012

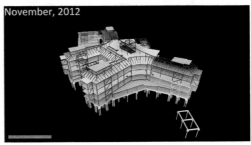

November, 2012

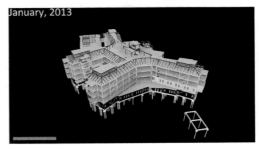

January, 2013

Figure 2.1.14
A sequence of still frames from a 4-D simulation

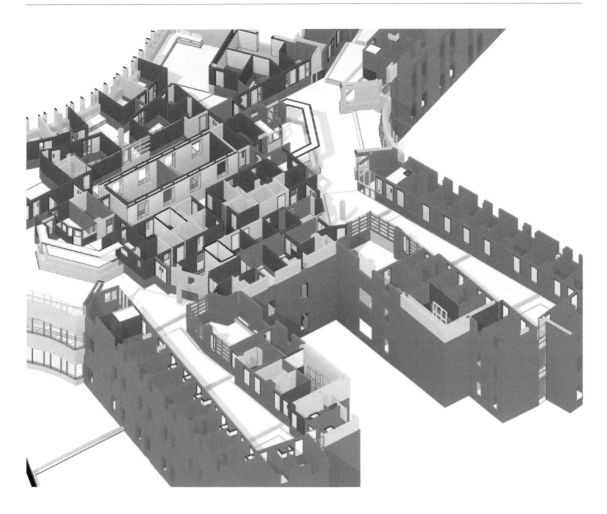

Figure 2.1.15
Construction Tracking

VDC-Based Progress Planning and Tracking
Tracking construction progress is essential to validate where a project is in relationship to the construction schedule, as well as determining what payments can be issued to the contractor. VDC tools allow an inspector to use the model to establish the completion status of each element, then use the software to color-code the status and report which quantities are complete. The information model is not only a powerful tool for generating progress reports, but the model itself becomes a recorded document of the complete sequence of a project. This can be extremely valuable after construction is complete if there are any disputes about sequential conflicts.

Model Quality Control

Models, data, and documents produced by the project team must be closely monitored to ensure coordinated data is generated for project execution. Knowledgeable 3-D operators are key to holding successful 3-D coordination meetings to identify and resolve design and construction issues, and for developing project-specific tracking procedures for the management and resolution of coordination tasks.

SUPPORT SERVICES

Risk Support

On large, complex projects, risk workshops are often used in order to gather all responsible parties in one room to analyze potential issues. But it is often very difficult to visualize where the potential issues lie. Information models are useful for visualizing multiple scenarios. 4-D schedules and their potential outcomes, configured through the information model, can be showcased live on-screen during meetings. This visual feedback clarifies different risk factors and dramatically improves the decision-making process, especially in the case of large stakeholder meetings.

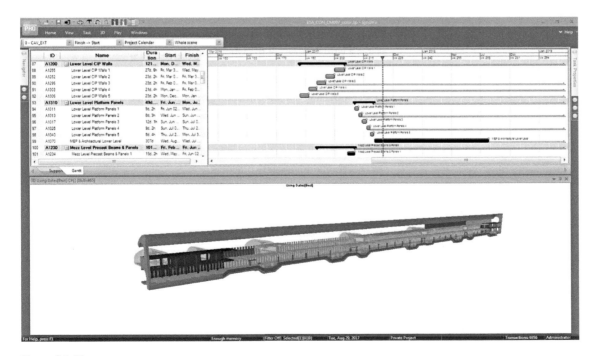

Figure 2.1.16
A screen shot of model in the Synchro software, which is often used for risk management support

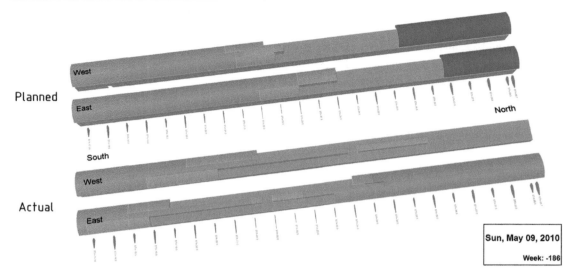

Figure 2.1.17
An image exported from an information model used for litigation support. The image shows planned and actual progress of a tunneling project

Claims Analysis and Mitigation

In today's complex construction industry, characterized by rigid time constraints, multiple prime contracts, and phased construction programs, claims have become a constant part of the construction process.
An information model can potentially help avoid possible litigation.
In case of litigation, the information model can provide simulations of the situation to clearly communicate the cause of the claim to all parties.
This is especially true if the project was tracked using a tracking model, as it is a virtual record of the whole construction project.

Worker Training and Safety

Project modeling and construction simulation can be used for simulating, planning, and demonstrating proper working methods to on-site staff. Utilizing a logistics model helps in visualizing and addressing safety concerns as well as supporting efficient on-site construction operations with comprehensive safety practices.

Construction Quality Assurance

Detailed on-site digital surveying and measurement technologies can assist with establishing actual grading geometry, site-built construction, and installation. On-site data can be compared against required locations of construction as specified in a project model. Comparisons of specific versus actual construction can be used to verify quality of construction, and support the accuracy of subsequent systems.

NOTE

1   These documents can be accessed through the following sources:
      www.nationalbimstandard.org and bimforum.org

2.2 Information Models

Information models are the cornerstone of successful VDC workflows. A variety of software platforms may be used to generate and manage the geometrical data for an information model, depending on specific project needs and team capabilities. Sufficient expertise is required for translating between different software platforms, generating appropriate deliverables, and for integrating and coordinating 3-D data generated by all members of the design and construction team.

Figure 2.2.1
Rendering of an information model exported from Revit

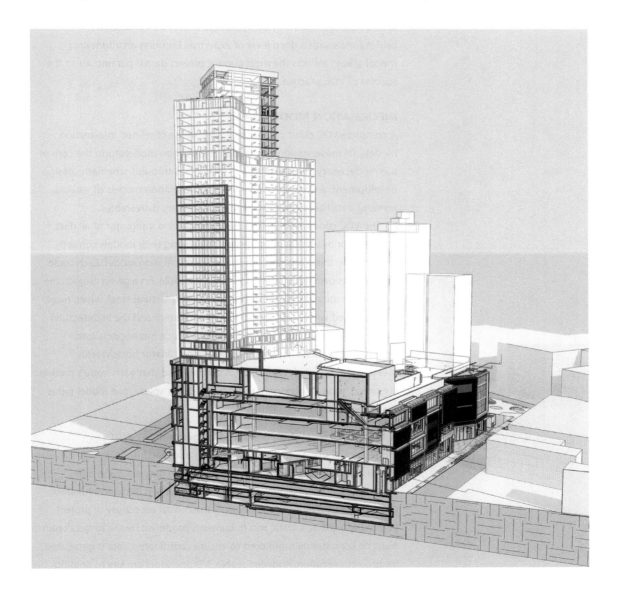

Effective modeling goes beyond integrating project data. VDC modelers should be integrated with the project team and understand the specifics of the team's specializations and workflows. Outsourcing information model authoring to outside firms, often in other countries, has become an increasing trend. However, this practice can be problematic as model authoring is an integrated process. Differences in expertise, time zones, and linguistics can often greatly complicate the collaborative process of efficient modeling. Construction practices vary significantly by region, and non-local modelers lack key knowledge of local building codes and construction processes. Especially during the construction phase, fabrication level systems models should be executed by professionals with a deep level of expertise. Ensuring an information model always reflects the most current project data is paramount to the success of VDC practice.

INFORMATION MODEL TYPES

A complete VDC effort comprises a wide range of related information models. Of these, design intent and construction models form the core of the model ecosystem. As a project progresses through schematic, design development, and construction phases, information models of varying levels of detail are needed to produce necessary deliverables.

The VDC coordinator acts as the manager and integrator of all data, ensuring various building trades are developing their models correctly, and unifying these models into the main information model. Each trade produces its own fabrication model. For example, on a given project, the structural engineer may produce a model of structural steel, which must be integrated with the model of the HVAC system and the architectural facade. A full-service VDC team can also produce more specialized information models as needed. The VDC coordinator hosts weekly meetings to ensure development is on track and that each trade's models are progressing in a coordinated fashion. These specialized model types are outlined below.

A Note on Information Model Management

Proper model management is essential to a successful VDC process. Models must be carefully maintained to ensure they are up to date, accurate, and reflect all key design and construction decisions. In effect, the model replaces 2-D documents as the central repository of project information. Models, data, and documents produced by the project team must be consistently monitored to ensure coordinated data is generated for project execution. Knowledgeable 3-D operators are key to holding successful 3-D coordination meetings to identify and resolve design and construction issues, and for developing project-specific tracking procedures for the management and resolution of coordination tasks. Typically one or more VDC coordinators assume these responsibilities, depending on the size of the project or scope of the VDC effort.

Design Intent Models

Design Intent Model

During the design phase of a project, the design team develops a design intent model, which is then updated during the construction phase when sketches and other design changes arise. In the traditional building process, language and information contained in the design intent model correspond to the SD, DD, and CD phase documents, accurately conveying the project's overall geometry with regard to structure, systems, and architecture. The purpose of the design intent model is to provide the design team with a tool to visualize and understand the integration of all of the building systems in 3-D space. The model is usually broken up by different design trades, such as Architecture, Structure, Mechanical, Electrical, Plumbing & Fire Protection, Communication and other special services. On larger projects, models can be further broken down by areas or subsystems; for example, within an Architecture model, the Exterior, Interior, and Vertical circulation might be separately linked models.

The design intent model matures along with the building process, reflecting the latest up-to-date information. By the Level of Development (LOD) standards, the schematic is considered LOD 100, Design Development LOD 200, and Construction Documents LOD 300. Design intent models reflect overall layout and basic sizing, but do not include specific fittings and hangers. The latest version of the BIM LOD standard has introduced LOD level 350, which includes enough detail to run effective clash detection.

Energy and Performance Models

A building's performance is measured mainly by its ability to provide comfortable conditions to its inhabitants and by the amount of energy it consumes. Decisions made by the design team are reflected in the performance of the building. The design intent information model can be analyzed for energy and building performance criteria and can be used to run a range of analysis simulations. This enables the design team to measure design decisions against both qualitative and quantitative measurements.

Construction Models

Trade Construction Models

Construction models are developed to a greater level of detail than the design intent model, corresponding to LOD 400. Developed by the contractor and the subcontractors, construction models are made to represent what is actually installed, as they form the basis for construction and fabrication. Each trade is responsible for creating its own trade construction model. Due to their high level of detail, trade models are usually subdivided into separate models for each floor in order to keep the files down to a manageable size. Furthermore, models need to be broken up because most fabrication software is still CAD-based and has no notion of levels. Contractors are responsible for aggregating the models and running coordination meetings to make sure all the different trades are in sync.

Survey Model

Creating a survey model is helpful for projects that relate to existing conditions. Having accurate geometrical information is critical in order to catch clearance issues and inconsistencies with the proposed design solution. A copy of this model will become the basis for the construction model. If no prior model exists, it is ideally modeled from existing 2-D drawings and then verified from a 3-D scanned point cloud. Point clouds can be linked into the authoring software, so the geometry can be modeled or adjusted to represent the true existing conditions. Several tools are now available that automate some of the conversion from the point cloud to BIM geometry. Pipes and topography generation in particular can benefit from this. It is important to make sure the correct coordinate system is determined, so the location of new models will be consistent. Establishing both the global and local (project) coordinate system is necessary at this point.

Layout Models

A construction model can also be used to generate layout models, which in turn drive robotic total stations layout machines. The information model requires special preparation for this purpose. Special point layout software helps automate the generation of these layouts. Points from the layout model can then be wirelessly sent to the layout machine, which then finds the correct point of layout in the physical space. Layout machines require only one person to operate and ensure that components are correctly installed as planned per the coordinated model.

Other VDC Models

Experience Simulation Models

To depict the experience of a proposed facility, information models can be exported into various visualization software, in which materials, lighting, signage, and other graphics are layered onto the model. Non-project-related objects, such as people, plants, vehicles, and other things that would populate a facility can also be added to the model. Camera paths are defined and conditions like time of day and weather can be simulated. Some objects can even be animated in order to convey a greater sense of place.

Coordination Models

A coordination model is a composite in which all of the models developed by different trades are linked. During the design phase, the architect operates a coordination model composed of the various design intent trade models. During the construction phase, the contractor manages a coordination model that combines the construction trade models. A coordination model is typically hosted in coordination software, such as Naviswork or Tekla BIMsight. More recently, cloud-based collaboration software, like the Autodesk 360 Glue platform and Assembly, are emerging as alternatives. Cloud-based software will eventually become the preferred technology, as it enables more advanced collaboration and easy access to the latest models, and requires minimal IT resources.

There are several benefits to using specialized coordination software. The software is able to read a wide range of file formats, and file sizes are more compact compared to authoring files, and therefore faster to run and easier to transfer. Geometry and data are locked down, so no one can accidentally alter the models. A coordination model is also continuously revised as consultants submit more developed models, and is used throughout the construction coordination phase of a project, primarily for clash detection between systems and constructability review.

4-D Models

The construction of a building is a sequential effort that requires detailed planning in order to understand the duration and sequence of a project. Project planning is done by creating a construction schedule that conveys all the various activities in a project over a period of time. A schedule can contain many thousands of activities. A Gantt chart helps plot out the sequencing of events, but it does not effectively communicate where in the project an activity is actually taking place and to what quantity. Often the only person that truly understands the schedule is the scheduler who created it. By linking activities to corresponding geometry in the information model, it is possible to visualize the schedule within the building and gain a much better understanding of how the project should come together. Types of activities and trades can be color-coded for effective visualization. A 4-D model is a 3-D model with the addition of the fourth dimension, time. 4-D models can be used like animations to understand the sequence of installation. They are useful both for planning and tracking work and comparing planned work to actual installation. A project schedule can be generated directly from a 4-D model using specialized 4-D scheduling software, such as Synchro.

Project schedules are reviewed in conjunction with corresponding spatial information to detect logistical conflicts or inefficiencies, and to support site management and planning activities. During early project phases, 4-D modeling can be used to develop, visualize, and analyze macro-level construction phasing strategies. Project modeling can be used as a basis for verifying certificates of payment for work installed. A 4-D model can represent a snapshot of a contracted plan for delivery of work. Additionally, throughout varying project phases, a 4-D model can be compared against actual site conditions to validate that work is on time and complete.

Tracking Models

As a project enters the construction phase, site work and construction progress can be captured in a tracking model. This model is used to visualize construction progress and becomes a record of the sequence of construction as it proceeds on-site. It is best to use a construction model as the basis for a tracking model, but tracking models need to be specially prepared for tracking purposes. Some geometrical elements must be further subdivided to reflect the way certain elements are actually installed. For example, a single slab might be poured in multiple pours, so the model would need to be subdivided into multiple sections to reflect this. A tracking model also needs to contain parameters that are designed to accept information coming from the field. Best practice would be to work closely with inspectors so the tracking model is updated in tandem with inspections. This way, it always contains the most up-to-date information regarding the state of progress on-site. As the information model contains quantities, it is possible to track completion ratios in order to drive payment requisitions. Technology for capturing field data is rapidly evolving, and there are specialized applications field personnel can use to collect data from the construction site. Cloud-based software such as 360 Field utilizes the data from the model to track equipment.

Cost and Quantification Models

Quantities derived directly from a 3-D model are usually more precise than conventional methods of estimation and they can be linked directly to cost data. This information is useful for project cost planning and project bidding.

Cost models are models in which quantity information is connected to quantity calculations and then linked to a cost estimation database. If the quantity and cost database are configured correctly, this drastically reduces the time it takes to produce a cost estimate. Changes to the schedule and the project and their impact on cost can be tested more easily. The Uniformat assembly system is often used to semi-automatically link the correct model component to the cost estimate. A thorough review of the model and its level of development must be conducted to ensure objects in the model are correctly defined and can be estimated accurately. Software like Innovaya can be used to help identify potential problems in the model. If the model-derived cost data is linked to a 4-D schedule, cost can be tracked throughout the lifespan of a project. This is often called a 5-D model.

As-Built Models

An as-built model is the final model produced and signed off by the project team. It is based on the actual installed dimensions and data. There can sometimes be quite a discrepancy between the geometry in the original design intent model and what is actually installed in the field. The as-built model assists in the effort to verify and document the project to ensure that the final drawings correspond with reality. With BIM-driven layout tools, such as the new robotic total stations, it is possible to achieve a higher level of accuracy of the installed systems as well as ongoing verification of the location of the installation. The documentation of the as-built model should be an integrated part of the installation process. This process is preferable to creating an as-built model after a job is complete, when many portions of the buildings are sealed off and not accessible for verification. This model also forms the basis for the facilities management model.

Facilities Management Models and Operations Models

Managing a facility requires knowledge of all operational systems of a building. To assist with this occupation, there are a range of Facilities Management (FM) and Operations/Maintenance (OM) software platforms. They are essentially powerful databases through which one can track and manage operations. Upon handover of a construction project, all necessary documentation and information has traditionally been put into binders, or saved in PDF format for the FM personnel to manually input into FM/OM software. As an information model is essentially a spatial database, it is the perfect vehicle for hosting and transferring critical data needed for operations into operations software. Techniques and industry standards are currently being developed for managing the migration of data from an information model to FM/OM software platforms. Some FM/OM software offers bidirectional links, so the information model can be used as an active FM/OM tool. Other software only accepts imports from information models.

3 Case Studies

Since its inception in 2010, LiRo's VDC department has worked on more than 15 projects, implementing VDC services to varying degrees. The projects, all located in the New York State region, have a combined contract sum of approximately $13 billion. Some of these have been executed as part of LiRo's construction management department, while on others the VDC department has been hired directly by the client.

To fully grasp how VDC services relate to AEC, it's crucial to examine how VDC has been effectively applied to actual projects. These case studies are included as means for exploring different VDC services. Each case study focuses on a particular aspect of the VDC process, and together the studies provide a full overview of the possibilities of VDC. They include in-depth discussions of systems coordination and design, project visualization, 4-D scheduling, logistics and constructability coordination, VDC construction management, tracking and reporting of site work, civil and structural model authoring, model review, and model integration.

This chapter begins with a discussion with LiRo Group CEO Lou Tormenta about the history and mission of LiRo's VDC department, how LiRo thinks about VDC and the firm's strategy for positioning itself as a technology leader in the AEC industry. The remainder of the chapter is devoted to case studies on various projects that the LiRo VDC department worked on between 2010 and 2014.

The case studies include the ground-up building and infrastructure projects of the Bronx Hospital of Mental Health, a $70 million publicly funded hospital in the Bronx; City Point, a $350 million mixed-use commercial and residential tower in downtown Brooklyn; and East Side Access, a $10 billion underground railway and train station connecting Manhattan and Queens. This chapter also includes a renovation case study: the RFK Bridge Toll Plaza, a $110 million infrastructure upgrade to the Robert F. Kennedy Triborough Bridge.

3.1 Interview with Luis Tormenta, CEO of LiRo Group

Luis Tormenta is a highly accomplished leader in the New York regions design and construction industry. He has been responsible for major public capital programs and has led many of New York City's most complex construction projects. As the founding commissioner of the New York City Department of Design and Construction, Tormenta managed the "super construction agency" created by then Mayor Rudolph Giuliani. As CEO, he is responsible for the overall strategic direction and management of the LiRo Group, with a focus on maintaining high-quality service standards, while increasing market share and expanding services nationally. Under his leadership, LiRo has expanded its service platforms to include disaster response management and virtual design and construction. He has also directed the company's national expansion, opening offices in Connecticut, Pennsylvania, and California. He is a board member of the New York Building Congress and the American Society of Military Engineers, and an adjunct professor in the graduate program for construction management at New York University-Polytechnic Institute. He holds a civil engineering degree from Manhattan College.

This interview was conducted by Lennart Andersson, Director of the Virtual Design and Construction group at LiRo.

DISCUSSION OF VIRTUAL DESIGN AND CONSTRUCTION

L. A.: Tell me about LiRo and its relationship to VDC.

L. T.: We have had the VDC department for about four years. About a year before we started the VDC group, there was a lot of talk about VDC but it was not very well defined. I had also read a little bit about it. For a company like ours, a CM and engineering business, it's important to be competitive in the marketplace. We always need to be looking at what we can be providing services-wise to our clients, not necessarily different services, but services that support the core of what we do. Sometimes, as has happened here, things take on a life of their own. Originally the idea was, "How can we use this technology to help our architects, engineers, and construction managers to do their job?" It was more from an internal corporate perspective, and I expected it to be more of a support role. It turned into something that helped us look at how we could do things in new ways, and, on the business end, it has helped us tremendously in terms of the work that we have won. That continues to be our focus, aside from winning specific VDC contracts that don't directly relate to what LiRo's other departments are doing, the competitive edge from a corporate perspective is an important part of what you guys do.

From day one, I found it to be really interesting. We started looking at what other people were doing, and we were asking ourselves whether maybe we should go out and hire somebody that could pick this up on their own and try to develop it. Which was the birth of our VDC department. I think from a business perspective, it's been really good for us: it gives us an edge. Some of the bigger firms also use VDC, but I haven't seen anything that seems to be as advanced, or that they handle it as comprehensively as we are trying to. I think there still are areas of improvement. How do we make it a valuable tool that can touch multiple different points in the services that we deliver? We now have a great modeling tool, and have developed ways to effectively tie the schedule to the model, the estimating function is now working. What about the possibility of connecting it to facilities management, how do we do that? Should we try to get into that service as a company?

Usually you get into these kinds of services, not necessarily because of a real strong strategic approach. Most of the time you have a strategic approach, and sometimes something falls next to you and you look over and say, "I think I can make something out of that," and that might work. But I think that Virtual Design and Construction as a tool to our business has been opening up new service possibilities.

Aside from CAD, what is the technology that has been traditionally applied in our field? Not on the materials side, where there has been great development, but on the methods side of our business—it's always been the same. I think the Romans did it much the same way that we do it today.

L. A.: In industries such as aircraft manufacturing, a lot of these integrated VDC technologies have been around for as long as 30 years. Why do you think that it has taken so long to reach the building industry? Is it the people?

L. T.: I think there is a sense that, for the most part, it relates to the complexity of the building compared to something like aviation or an industrial plan. Part of it is that it's a relatively simple process to build a building, or build infrastructure, do we really need the high-end tools? That was the mentality. I think it's also about learning the technology and the fear factor. If you go to a university and your major is in aviation structural design or an industrial process, you come out of it with a mentality that's much more accepting of advanced technology, as you are working to build something very complex, as opposed to a lot of buildings.

I think people are reluctant to change their perspective of, "Do I really need to learn it?" There is a generational gap. At my age, when I graduated from engineering school, we did not have desktop computers, we were using Fortran computing punchcards that were loaded on a mainframe. And if something was wrong, you had to go through thousands of lines of code to correct it. You have people like me, my generation in my mid-fifties, who are the people that decide to what extent this technology should be used. Because we did not grow up with it, there is such a big generational gap. People of my generation will walk off into the sunset, and the people that will replace them grew up with this kind of technology as second nature. Things like cell phones and other mobile technologies.

> L. A.: My observation is the new generation often think they wouldn't be able to work without using these tools.

L. T.: Back in school we drew everything by hand and used a Texas Instrument calculator that cost $400 and could do simple calculations. That was the technology of the day. I think as the current generation grows, the technology will become more and more integrated into everyday operations. It becomes part of the fabric, as opposed to a guy like me who didn't grow up with it.

> L. A.: Considering the long-term outlook, do you think the LiRo's VDC team will slowly become integrated into the larger company, or will it continue as a separate department within LiRo?

L. T.: From my management methodology, the VDC group needs to continue to exist as its own separate entity. It needs to champion the latest technology, to push the latest ideas without being hindered by the production and everyday management of projects. What I think what will happen is the next generation of project managers will require VDC and ask for the latest tech, which the VDC group can provide. However, to continue to develop and integrate these evolving tools, there will always be need for this group.

> L. A.: With VDC, there are many promises as to how it can save money and time as well as deliver other benefits to projects. Do you think these are possible to measure?

L. T.: I have always been interested, from a management perspective in our business, in what every client wants: They want a quality job, on time and on budget. We go into these presentations and say, "We are going to do that." I know from years of experience that there are very few projects where all three of these criteria are met. So then the question is: How do we approach these projects, not only construction projects but any project? What is the secret, the holy grail to actually make these projects happen?

When we start analyzing, it's sort of like a recipe—like you are going to cook a dish. What are the critical ingredients that go into the dish that you are cooking? And if any of those get screwed up, how will the dish taste?

So if you look at a project, you have the client and the client representative that has decisions to make; they have to be an educated client because they have to make those decisions in a timely way. The construction manager needs to understand how to move the project along and what the processes are, the designers hopefully have done a quality job in transforming the ideas and concepts from the client's mind to documents. There will always be some discrepancies, but not too many so there is at least something to follow. Then there is the contractor that is actually going to execute. Who of those parties has the potential to have the greatest impact on these three items? Which one of those and at what point of time on the project and how do you measure that? Going back to your question, I read an article about this study, which plotted out thousands of projects of the reasons for delays.

If you establish a bell curve of the standards of a whole series of projects, has BIM moved this bell curve? What are the averages for change orders? You can put a price on the change orders and see statistically if they are reduced. However, it is hard to measure as every project tends to be so different.

I am noticing that the true value of VDC becomes clear when everybody is on board. There are many levels of VDC, but to achieve a major impact on performance, it has to be integrated and all participants have to do their "VDC part."

The process of coming on board is primarily about education. How many contractors do you know that are going to buy into it? Especially if they are not forced into it, or unless they clearly realize that they are going to get a financial benefit out of it. There are certain contractors and trades that have done that. The sheet metal guy is clearly benefiting from modeling up their trades, especially compared to their old ways of drafting the complex shapes with pencil.

L. A.: Are VDC services changing how LiRo is conducting CM services?

L. T.: I think it can help us deliver what we truly expect to deliver. When we say that we are going to do something, the level of certainty or guarantee that comes along with that, VDC can really help us to be more accurate. I think it is a really sharp tool to help us get a higher degree of success. It's like the surgeon who will be able to use a laser instead of a scalpel when halfway through to get the tumor in the brain, and he erases half the patient's memory. OK, that is not acceptable, but if that is all you have when you need to get rid of the tumor, you do it. But if you have better, more precise tools you will of course use those. So VDC will help us to become much more accurate in what we do.

L. A.: Hype and promise. What are the cons or negatives of VDC?

L.T.: If we get too dependent on that alone. If I think I don't need to hire a really experienced designer or CM, because I can hire a bunch of kids out of school making $40,000 a year. And because we are using these high-tech tools, they will be able to design, detail, and coordinate because the tools will be doing it for them. They don't need that experience: That is the danger. The tool does not do it for me. We are losing drawing skill. We can get too dependent on the tool doing the thinking and not fully understand what we are working with.

L. A..: Your previous expectations of VDC—has it delivered?

L. T.: I didn't know much about it—I knew that it was something we probably should get ourselves into, as I said earlier. I think there hasn't yet been total buy-in by all the parties that make up a process, so until that happens, I don't think we have seen the full potential yet. It's not a technology issue anymore, but how you get people to start using it. How do we use it collaboratively, so that everybody benefits from it? It's still a little here and a little there, but yet not fully integrated in full execution of the process. That is where I think it is right now—it is just beginning to truly affect the industry.

3.2 City Point Phase II

City Point is a mixed-use development in downtown Brooklyn off Fulton Street, the borough's main retail corridor. Once completed, City Point will be 1.6–1.9 million square feet and include retail, affordable, and market rate housing as well as a new park. At the time of writing, the first phase, a building consisting of 50,000 square feet of retail space, has been completed. The second phase is a composite of three projects: a five-storey retail podium with two separate residential towers above it. Tower one is 35 floors of affordable housing and tower two is a 45-floor market rate residential rental tower. The Phase 2 projects were designed and constructed concurrently by different developers, designers, and construction teams. A third tower of 65 storeys is currently in the planning stages for Phase 3 and will be completed in 2020. The LiRo VDC team was brought in to generate an information model from the 2-D CAD design documents and to help manage the complexity of the project.

Overall Project Budget:	$430 million
Location:	Downtown Brooklyn, NY
Completion:	2014–2016
Type:	Mixed use, commercial, residential
Clients:	Acadia Realty Trust, Brodsky Organization & Gensler
Fee Structure:	Lump sum for modeling and hourly for coordination
Contract Type:	Design, Bid, Build
VDC Focus:	Systems coordination and design; project visualization
VDC Technologies	
Information Model Authoring:	Autodesk Revit 2013
Coordination and Clash Detection:	Autodesk Navisworks 2014
Project Information Management:	Newforma
Visualization/Animation:	Lumion, Adobe Premiere

Figure 3.2.1
City Point under construction—
photograph facing south along
Gold Street

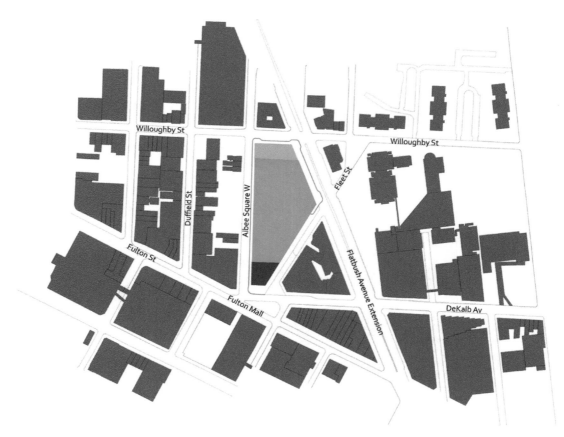

Figure 3.2.2
Project map

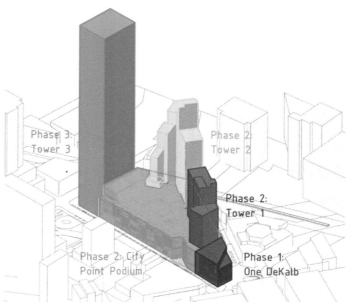

Figure 3.2.3
Project overview and phases diagram

SUCCESSES
- Identified major interferences and resolved serious clearance issues between the structure, architecture, and MEP systems.
- Created a solid foundation of information modeling during the design stage, which was key to implementing VDC during the construction phase. The design phase information model was given to the subcontractors to further develop into trade construction models.
- Created a high-quality experience simulation based on the information model, which provided the client with an innovative approach to marketing and sales.
- Identified all required beam penetrations in conjunction with the engineering team, minimizing the need for field-engineered penetrations and saving substantial costs to the client.

CHALLENGES
- Educating and guiding the subcontractor team, who had no prior experience with information modeling or VDC, through the VDC processes.
- Working with subcontractors who had outsourced modeling to low-cost firms located outside the United States, resulting in considerable delays to the coordination schedule.

PROJECT BACKGROUND

Coordinating building trades is a challenging operation, even on a fairly small project, and for City Point it was especially so. Three completely different project teams required concurrent cross-coordination. Tower two was a union project designed with a concrete structural system, while the retail podium employed non-union labor and was designed with a steel structural system. Acadia Realty Trust, the developer of the podium, initially engaged LiRo to create a design intent system model for coordination during the design phase. The goals were to uncover and resolve all building component interferences and validate retail space height requirements. The success of the Acadia Realty Trust's initial scope led to a contract with the Brodsky Organization, the developer for tower two, to model the most critical systems for the tower and work with their subcontractors to ensure proper coordination with the podium team. Another VDC consultant was separately contracted to model the tower one systems.

Gensler, the architect for the podium's anchor retail store, also hired LiRo's team to model its interior and systems for coordination with the podium's other building systems. The scope included modeling interior partitions, ceilings, light coves, and different supporting systems to accurately adhere to clearance requirements. The model revealed that the

Figure 3.2.4 (overleaf)
Timeline with products and services

PHASE

INFORMATION MODEL

VDC SERVICES

VDC PRODUCTS

CONCEPTUALIZATION

DESIGN

PROCUREMENT

PRE-CONSTRUCTION

Design Intent Model

Construction Management + VDC Models

Construction Models

Design Coordination

Experience Simulation

Distribution models

vRFI

Virtual Requests for Information

Tennant Coordination drawings

Walk-throughs/ Fly-throughs

CONSTRUCTION

OPERATION

DECOMMISIONING

Construction
Coordination

 Distribution
models

 Virtual
vRFI Requests for
Information

 Clash
Reports

 Systems
Coordination
drawings

CITY POINT ORGANIZATIONAL CHART

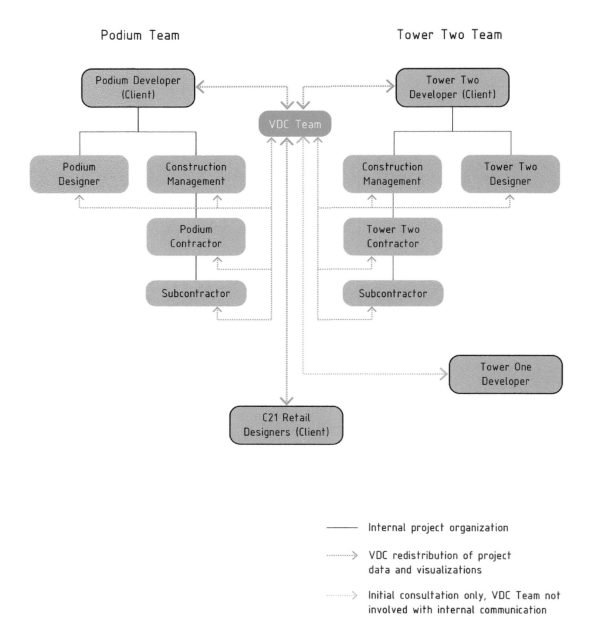

Figure 3.2.5
Diagram showing City Point project team organization

retail ceilings needed to be lowered to accommodate the systems. Mechanical rooms were also added to serve the retail space, which required some redesign of the podium's base building systems. The integrated review process between the base building, tower, and retail tenant made it possible to optimize the design and guarantee maximum ceiling heights.

The Coordination Model generated during the design phase was transferred to the contractor at the beginning of the construction phase for subcontractors to use as a guide for modeling their systems. LiRo VDC ran coordination meetings throughout the design and construction phases, and took an active role in guiding the design team and subcontractors through the coordination process.

Design Phase

Initially, Acadia Realty Trust was planning to model the architecture in-house and have LiRo model the MEP systems. However, Acadia had limited experience with information modeling, and when they realized the complexity of the architectural modeling, they added architecture to LiRo's scope of work. LiRo's team joined the project just as the 80 percent construction documents were issued. As construction coordination progressed, 12 additional bulletins were issued, resulting in additional time spent reviewing and updating the information model. Some of these

Figure 3.2.6
Perspective view of Century 21 systems coordinated with Phase 1 and Phase 2 systems. The different contracts were color-coded: Red = Podium; Purple = One DeKalb (Phase 1; Blue = Tower One; Green = C21)

revisions were the result of issues found during the coordination process. In an ongoing coordination process, it is difficult to forecast the complexity of the project as well as the skill sets of various parties.

The model also became a valuable visualization tool. The developer asked for signage, store branding, and specified materials to be added to the design intent model to generate near photo-realistic walk-throughs of the project. The resulting six-minute walk-through animation showcased the building's surroundings, exterior, and retail interiors. The animation was brought to life with retail models, cars, vegetation, and people. Turning the animations around quickly made it possible to incorporate design changes and generate multiple versions customized to potential retailers.

The design intent model was used only for reference purposes to find discrepancies and visualize issues. Design documents generated by the AE team were the legally binding documents used for the bidding process. Modeling was done in the LiRo office, while coordination meetings were held in the field office. These meetings included the architect, the MEP, and structural engineers as well as the developer. Two separate large displays were used in meetings, one showing the Navisworks 3-D coordination model, and the other showing 2-D construction document floor plans in Revit. These plans showed both the modeled systems as well as the 2-D drafted construction documents. Plans are the most effective way to quickly locate the area under discussion and orient the team to review the 3-D information model.

Construction Phase

At the commencement of construction coordination, a kickoff meeting was held to introduce the subcontractors to VDC. An implementation plan that contained information about essential standards and procedures was distributed during the meeting. The subcontractors were responsible for modeling their systems to a construction level of development (LOD 400), while the VDC team was responsible for compiling 3-D models from all of the subcontractors into the coordination model and conducting interference reviews.

To ensure smooth coordination, LiRo recommended that each subcontractor hire an in-house modeler who would be available to attend coordination meetings and update and review their respective model. Unfortunately, some of the subcontractors hired modeling consultants who outsourced modeling tasks because they did not have personnel with adequate skill sets in-house. This was detrimental to the coordination process, as the outsourced modelers did not have local building

VDC PRODUCTS

Design Intent Model An LOD 300 model which includes architecture and structure as depicted in the construction documents.

Coordination Model The central information model that holds the structural, architectural, and systems sub-models. Managed by the LiRo Virtual Construction Coordinator and used to generate vRFIs and clash detection reports.

Construction Model(s) Models for each system to include all elements, equipment, fittings, etc., accurate in size, shape, location, quantity, and orientation with complete fabrication assembly and detailing info as required in LOD 400 (AIA-document E202). Each model is broken down by subsystem for each level.

Distribution Model The current Navisworks version of the coordination model distributed before each coordination meeting. Contains all of the clash views.

Systems Coordination Drawing Set The systems coordination drawing set consists of floor plans generated by level and by system from the coordination model. Plans contain the 100 percent CD CAD set with the design intent model and the construction models overlaid.

Tenant Coordination Drawings This drawing set was generated from the model in order to provide underside elevations of the system runs and guarantee clearance requirements. The retail tenants had specific requirements for clearances, and this set ensured there would be no unexpected changes from the original assumptions when tenants finalized store layouts.

Shared Folders Folders within the Newforma Project System for subcontractors to upload construction models.

Clash and Interference Reports Reports generated from the coordination model detailing specific clashes between the systems, architecture, and structure models. Also saved as views in the coordination model.

Experience Simulation Movie Detailed walk-through animation of the exterior and key interior spaces to convey design changes and ideas; also used for marketing purposes.

VDC SERVICES

Architectural and MEP Systems Modeling All architecture and MEP systems were modeled and organized to facilitate effective coordination.

VDC Implementation Plan The design and construction phase coordination process was outlined in a detailed VDC implementation plan.

Design Phase Trade Coordination The VDC team managed the coordination process of the systems with the design team.

Construction Phase MEP System Coordination During the initial phase of the construction coordination the VDC team managed the coordination of MEP systems.

experience and could not participate in the coordination meetings to resolve clashes in real time. It is imperative to the success of the coordination process that modelers be directly involved in resolving interferences with the subcontractors responsible for installation. Having outsourced modelers resulted in delays, as some areas had to be reworked multiple times.

IN-DEPTH DESCRIPTION OF SELECTED VDC SERVICES

Design Phase MEP Systems Modeling

MEP systems modeling is one of the most complex and time-consuming types of modeling. The intricate systems require substantial effort and an ample amount of time to coordinate in the limited space above the ceiling. Traditional contractual structure does not include compensation for spatial coordination; therefore the engineering community often perceives VDC as additional work for the same pay. Contractual structures need to evolve to include fair compensation in order for MEP engineers to embrace VDC. LiRo's MEP group has adopted a fee structure that includes additional compensation for projects that involve VDC coordination. They will not release the information model unless VDC services are explicitly incorporated into the contract. Modeling MEP systems has tremendous benefits for the outcome of a project. A common reason for change-orders during construction originate from a lack of coordination during the design phase. It is critical to have a clear, organized methodology for systems modeling. Doing this type of modeling in an undisciplined way can cause an enormous amount of rework as the project moves into construction.

The structural system and architectural components in the design intent model must be sufficiently developed before incorporating MEP systems. It is only possible to model these system elements within their architectural and structural context. As traditional 2-D MEP plans depict the system routes fairly diagrammatically and provide only generalized guidelines for routing elevations, placing elements at optimal heights requires constant analysis of the context. Effective MEP modeling is a skill set that requires deep knowledge of systems engineering, modeling tools, local building codes, and construction techniques. MEP engineers need to be knowledgeable about a system's access requirements, hanger types, seismic regulations, and installation sequence. They also need to be in constant dialogue with the design team to find the best routing for piping and ducts.

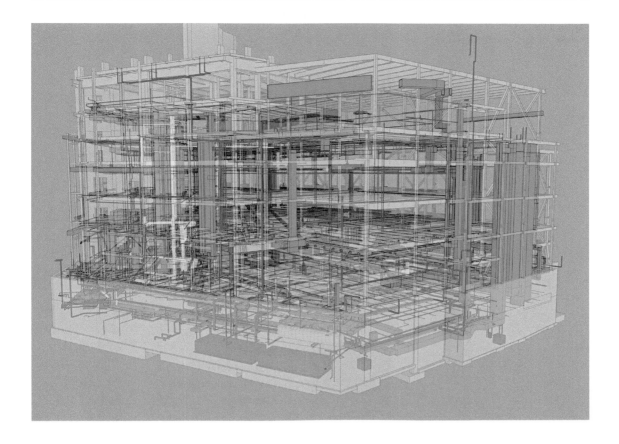

The modeling of pipe and duct runs should be sequenced in a logical way based on flexibility and dimension. The general rule is to model mechanical equipment and ducts first, as they are the largest elements, and points where ducts cross over each other are a common cause for insufficient clearances. Next in priority are sloped pipes, which can cause clearance issues at their lowest point, especially for long horizontal runs.

A clear and consistent color-coding system is essential to read the model and quickly understand which pipes and ducts belong to a given trade. LiRo VDC has developed two different types of color-coding systems. One assigns a color to each trade, so they can be distinguished from one another. The second color-codes the subsystems (such as return or supply ducts) for coordination within each trade. At City Point, a third system was sometimes used to distinguish between different contracts.

Figure 3.2.7
Image of podium systems

Figure 3.2.8
Diagram of color codes for systems

COLOR CODE #1		COLOR CODE #2
Trade+ Color	System	System Color
MEC	Return Air	
	Supply Air	
	Exhaust Air	
	Makeup Air	
	Outside Air	
	Chilled Water Return	
	Chilled Water Supply	
	Condensate Drain	
	Condenser Water Return	
	Condenser Water Supply	
	Heating Hot Water Return	
	Heating Hot Water Supply	
	Natural Gas	
	Other	
	Mechanical Equipment	
PLU	Domestic Cold Water	
	Domestic Hot Water	
	Domestic Hot Water Return	
	Ejector Discharge	
	Other	
	Sanitary (Blackwater)	
	Storm Water	
	Vent	
	Waste	
	Plumbing Fixtures / Eqt	
FPT	Fire Line	
	Sprinkler Line	
	Sprinklers/Fire Protection Eqt	
ELC	General Conduits	
	Electrical Equipment	
COM	General Conduits	
	Communication Equipment	

COLOR CODE #3	
Tower 1	
Tower 2	

Successful coordination drawings and models are organized, and therefore quick and easy to understand. During the design phase for the podium project, LiRo created all of the mechanical, electrical, and plumbing systems models for the engineers, so the whole system was modeled in one Revit file and sorted based on worksets. Because all of the systems were modeled in a single file, they could all be adjusted concurrently, and rerouting among multiple disciplines could be accomplished at once. The LiRo team linked in all the 2-D design drawings and printed PDF files of a complete coordination drawing set with the color-coded model on top, so any discrepancies could be easily spotted. There were so many changes made that eventually the MEP engineers signed off on the drawings created by the VDC team and issued them directly to the contractor. Once coordination was complete, all changes were incorporated into the design team's CAD documents and issued as a bulletin. The color-coding capabilities of information modeling greatly enhance effective communication, particularly in the case of MEP systems.

The tenant coordination drawing set was an additional product of design coordination that proved useful to the podium's retail tenants. This drawing set was generated from the model in order to provide underside elevations of the system runs. The retail tenants had specific requirements for clearances, and this set guaranteed there would be no unexpected changes in ceiling height from the original assumptions when tenants finalized store layouts.

Design Phase vs. Construction Phase Coordination

The design and construction trades have very different points of view. Designers tend to focus on the end result, while construction teams focus on means and methods—how something is assembled. When looking at a traditional design-bid-build project, there is often a huge disconnect between the two trades, as most of the design is done before any construction professionals are involved. Especially with regard to systems, contractors and installers have expert knowledge of how system components are fabricated and installed, and therefore drive how systems should be organized. The sequence of installation among trades can also influence how systems should be layered. During the design phase, it is only beneficial to resolve major issues with pipes and ducts. Hangers, wrappers, and fittings will be added during the construction phase, and their exact routings determined at that time.

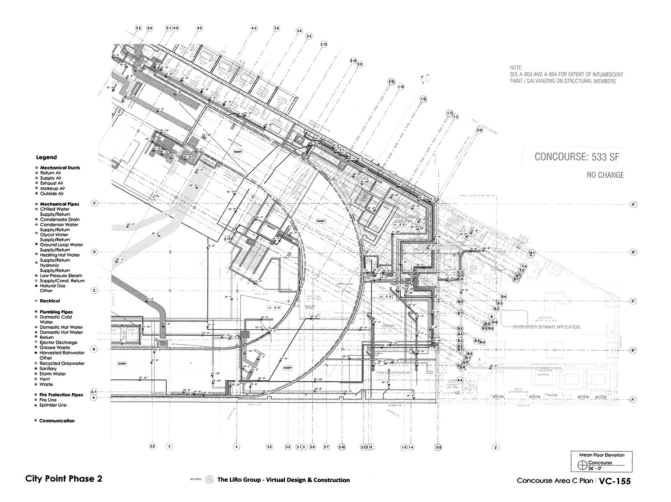

The LiRo Group - Virtual Design & Construction

Concourse Area C Plan | VC-155

Figure 3.2.9
Tenant coordination drawings showing elevations of
system runs

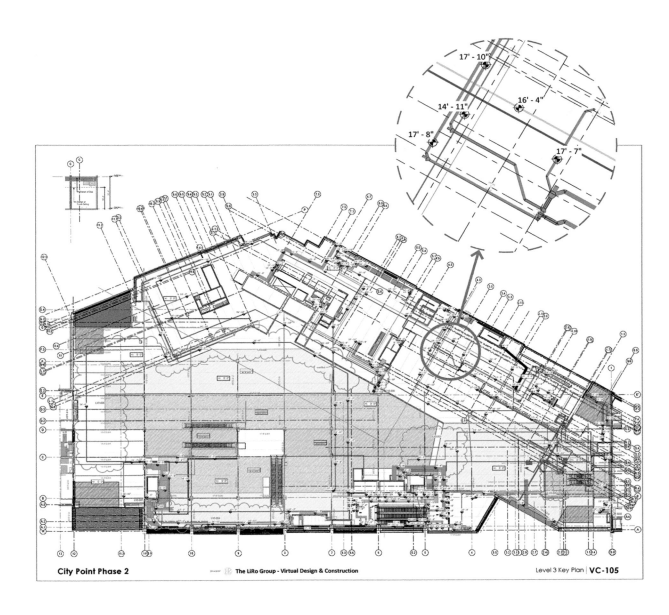

17' - 10"

14' - 11"

16' - 4"

17' - 8"

17' - 7"

City Point Phase 2

The LiRo Group - Virtual Design & Construction

Level 3 Key Plan | **VC-105**

CITY POINT DESIGN INTENT COORDINATION STRUCTURE

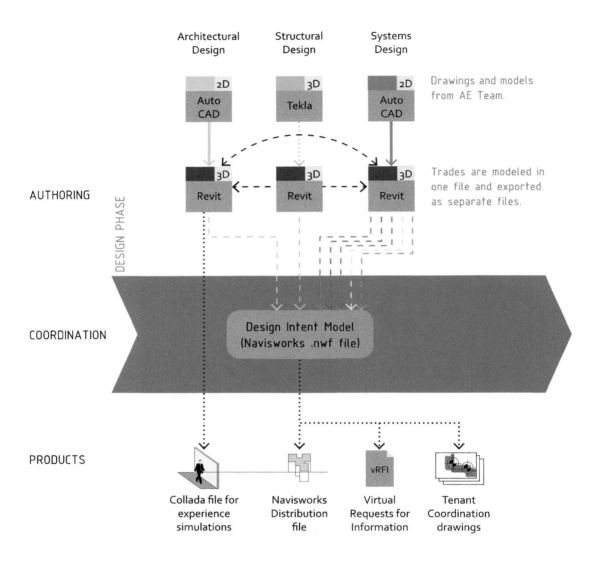

Architectural Design

Structural Design

Systems Design

AUTHORING

COORDINATION

PRODUCTS

DESIGN PHASE

Drawings and models from AE Team.

Trades are modeled in one file and exported as separate files.

Design Intent Model (Navisworks .nwf file)

Collada file for experience simulations

Navisworks Distribution file

Virtual Requests for Information

Tenant Coordination drawings

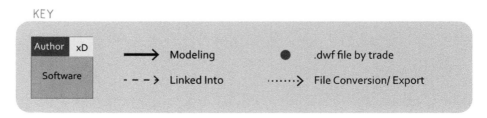

KEY

Author xD

Software

→ Modeling

- - -> Linked Into

● .dwf file by trade

·······> File Conversion/ Export

Figure 3.2.10
Diagrams of design intent coordination structure and construction coordination structure

CITY POINT CONSTRUCTION COORDINATION STRUCTURE

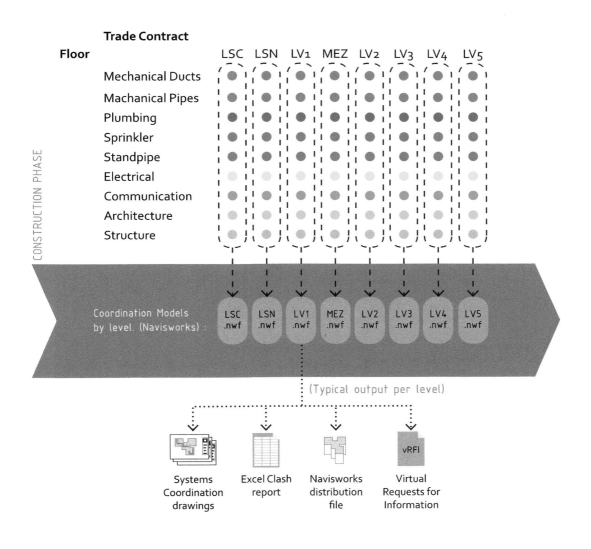

Trade Contract

Floor

CONSTRUCTION PHASE

	LSC	LSN	LV1	MEZ	LV2	LV3	LV4	LV5
Mechanical Ducts	●	●	●	●	●	●	●	●
Machanical Pipes		●	●	●	●	●	●	●
Plumbing	●	●	●	●	●	●	●	●
Sprinkler	●	●	●	●	●	●	●	●
Standpipe	●	●	●	●	●	●	●	●
Electrical	●	●	●	●	●	●	●	●
Communication	●	●	●	●	●	●	●	●
Architecture	●	●	●	●	●	●	●	●
Structure	●	●	●	●	●	●	●	●

Coordination Models by level. (Navisworks) :

LSC .nwf	LSN .nwf	LV1 .nwf	MEZ .nwf	LV2 .nwf	LV3 .nwf	LV4 .nwf	LV5 .nwf

(Typical output per level)

Systems Coordination drawings

Excel Clash report

Navisworks distribution file

vRFI
Virtual Requests for Information

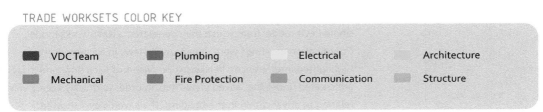

TRADE WORKSETS COLOR KEY

■ VDC Team	■ Plumbing	■ Electrical	■ Architecture
■ Mechanical	■ Fire Protection	■ Communication	■ Structure

CITY POINT SUBCONTRACTOR COORDINATION CLASH MATRIX

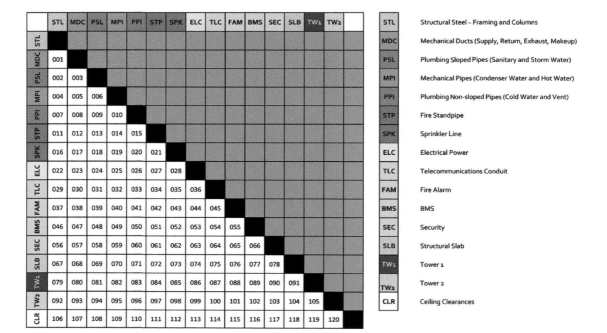

	STL	MDC	PSL	MPI	PPI	STP	SPK	ELC	TLC	FAM	BMS	SEC	SLB	TW1	TW2
STL															
MDC	001														
PSL	002	003													
MPI	004	005	006												
PPI	007	008	009	010											
STP	011	012	013	014	015										
SPK	016	017	018	019	020	021									
ELC	022	023	024	025	026	027	028								
TLC	029	030	031	032	033	034	035	036							
FAM	037	038	039	040	041	042	043	044	045						
BMS	046	047	048	049	050	051	052	053	054	055					
SEC	056	057	058	059	060	061	062	063	064	065	066				
SLB	067	068	069	070	071	072	073	074	075	076	077	078			
TW1	079	080	081	082	083	084	085	086	087	088	089	090	091		
TW2	092	093	094	095	096	097	098	099	100	101	102	103	104	105	
CLR	106	107	108	109	110	111	112	113	114	115	116	117	118	119	120

STL	Structural Steel – Framing and Columns
MDC	Mechanical Ducts (Supply, Return, Exhaust, Makeup)
PSL	Plumbing Sloped Pipes (Sanitary and Storm Water)
MPI	Mechanical Pipes (Condenser Water and Hot Water)
PPI	Plumbing Non-sloped Pipes (Cold Water and Vent)
STP	Fire Standpipe
SPK	Sprinkler Line
ELC	Electrical Power
TLC	Telecommunications Conduit
FAM	Fire Alarm
BMS	BMS
SEC	Security
SLB	Structural Slab
TW1	Tower 1
TW2	Tower 2
CLR	Ceiling Clearances

Figure 3.2.11
Construction clash detection matrix

Design Phase Coordination

LiRo was responsible for modeling the architectural and MEP aspects of the information model, while the structural model was created by the structural engineer using the software Tekla. For coordination, the structural model was provided to LiRo through the common model exchange format IFC, which was imported into Revit and linked to the architecture and MEP information models. The coordination process consists of eliminating major discrepancies and interferences between the architecture, structure, and MEP systems of the building. (Using the term "interferences" rather than "clashes" was a conscious choice, as many of the conflicts related to a host of problems that needed resolution, including clearances, best routing, access requirements, etc.) To run interference detection, the LiRo team compiled the three models into a single Revit model from which they generated a Navisworks model. Coordination meetings were driven using views in Navisworks that highlighted interferences. In Navisworks, views of each floor and location are created under the "saved views" tab, in order to quickly access views relating to any location in the building. The saved views showed contractors areas with interferences that required some resolution or further coordination.

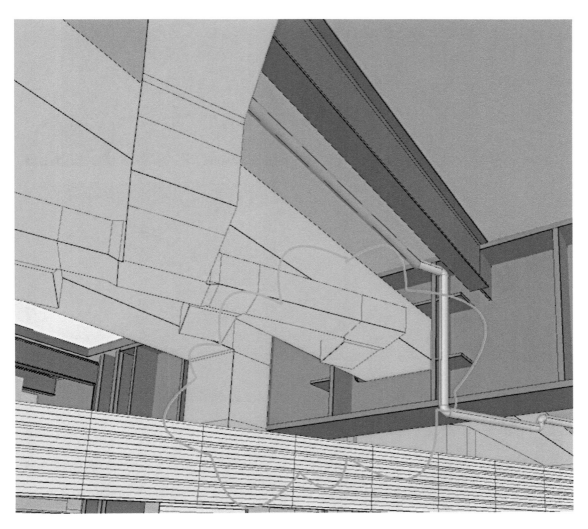

Figure 3.2.12A
Design phase coordination: identifying duct and beam interferences during the design phase minimized the need for field-engineered beam penetrations, saving substantial costs to the client

Figure 3.2.12B Further examples of interferences

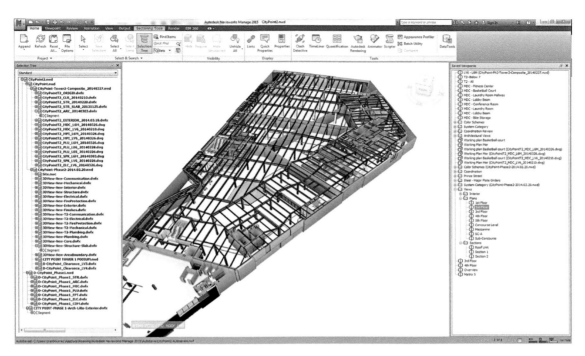

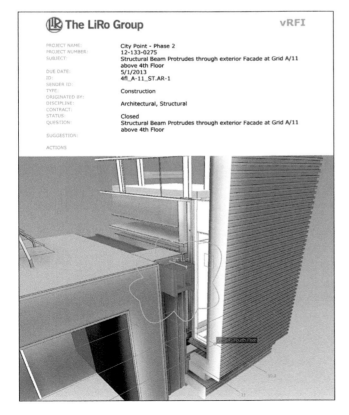

Figure 3.2.13
Views of each floor can quickly be accessed in Navisworks via the "saved views" tab. Here, the sub-concourse view shows the architecture, structure, and mechanical systems associated with this level

Figure 3.2.14
An example of a vRFI document

Chart showing number of City Point
Phase 2 clashes over time

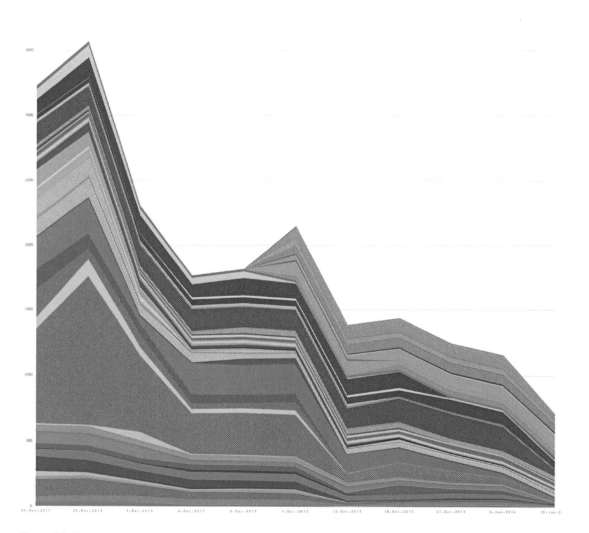

Figure 3.2.15
Chart showing interferences decreasing over time

Issues were divided into two categories: minor and major. Minor issues were those which could be solved with simple rerouting. Remodeling minor issues was not seen as an efficient use of time during the design process, as these could be more productively addressed during the construction coordination phase. Major issues were processed as Virtual Requests For Information (vRFIs) using a Project Information Management system called Newforma. The vRFIs were sent to the responsible parties with attached screenshots from the model. In this way, issues were automatically tracked and agendas generated for coordination meetings. Major issues were remodeled and fully resolved during the design phase.

Construction Phase Coordination

During the kick-off meeting for the City Point podium, subcontractors were provided with complete design intent models of the systems, structure, and architecture. As many subcontractors use CAD-based systems for modeling, it is essential to give them 3-D files that are compatible with their software. When construction phase coordination starts, the model is broken up into individual 3-D views of each level and exported as separate AutoCAD 3-D DWG models. Some subcontractors use Revit and can work directly from the Revit geometry, rather than from the exported DWG models. Construction phase coordination is most effective when the subcontractor models are created by an in-house modeler for each trade, as modeling for fabrication requires a deep knowledge of the system's technology, relevant building codes, cost of components, and other factors of installation and assembly.

The coordination proceeded with bi-weekly, daylong meetings to resolve major issues. The modelers for each trade were required to attend with their laptops and their individual working 3-D models. The LiRo VDC coordinator was responsible for leading the meeting and operating the coordination model—a composite Navisworks model including the podium's architecture model, structure model, and 3-D systems models from each of the subcontractor modelers. The architect and engineers were scheduled to attend portions of the meetings to address issues that required design input.

After the first few coordination meetings, during which issues were resolved pertaining to floor elevations, ceiling heights, and code requirements, LiRo VDC began the Navisworks clash detection process. The model was dissected level by level by the whole coordination team to identify and resolve all system clashes. Since there were 12 different subcontractors participating in the construction phase clash detection

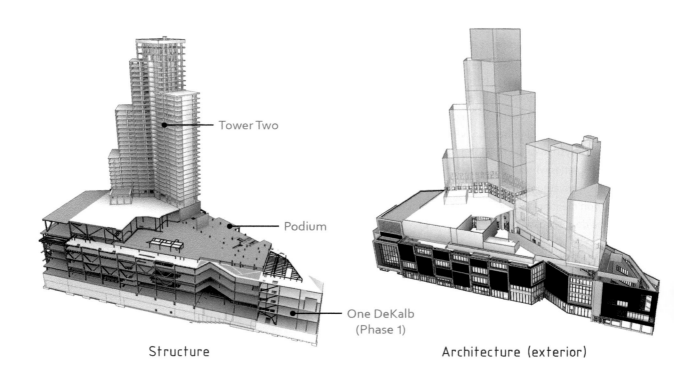

Tower Two

Podium

One DeKalb
(Phase 1)

Structure

Architecture (exterior)

Architecture selectively modeled
to suit scope of trade coordination
between Tower Two and Podium

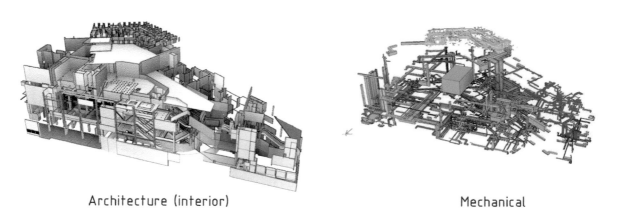

Architecture (interior)

Mechanical

Figure 3.2.16
The various models that make up the City Point composite design intent Navisworks model. The LiRo VDC coordinator was
responsible for operating this model, and scanning it for interferences

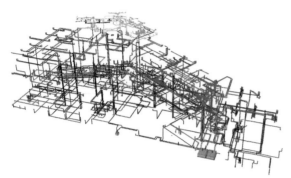

Plumbing

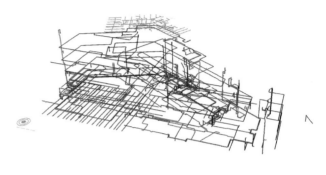

Fire Protection

Electrical

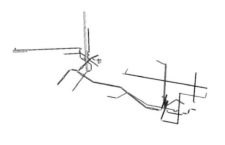

Communication

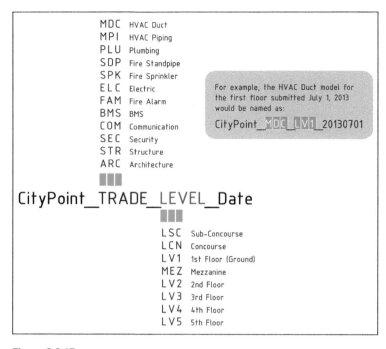

MDC   HVAC Duct
MPI   HVAC Piping
PLU   Plumbing
SDP   Fire Standpipe
SPK   Fire Sprinkler
ELC   Electric
FAM   Fire Alarm
BMS   BMS
COM   Communication
SEC   Security
STR   Structure
ARC   Architecture

For example, the HVAC Duct model for the first floor submitted July 1, 2013 would be named as:

CityPoint_MDC_LV1_20130701

CityPoint_TRADE_LEVEL_Date

LSC   Sub-Concourse
LCN   Concourse
LV1   1st Floor (Ground)
MEZ   Mezzanine
LV2   2nd Floor
LV3   3rd Floor
LV4   4th Floor
LV5   5th Floor

Figure 3.2.17
File-naming diagram

process, 12 trade models were generated for each of the building's eight floors. The resulting 96 separate models needed to be organized and continuously updated. A clear file-naming convention was used to keep the system organized: a three-digit format to describe the trade, level, and date.

The subcontractor modelers were required to have enough expertise to adjust their models in the meetings to resolve specific clashes, both from the standpoint of the technical operation of the model and familiarity with the system's requirements and rules. When a clash-free model (with no geometric clashes or other major issues) for the level was obtained, it required a sign-off from the construction manager's MEP superintendent for subcontractor coordination of that level to be concluded. As a result of this construction coordination effort, the VDC team was able to provide the entire construction team with a system coordination drawing set, consisting of floor plans generated by level and by system. Plans contained the 100 percent CD CAD set with the design intent model and the construction models overlaid.

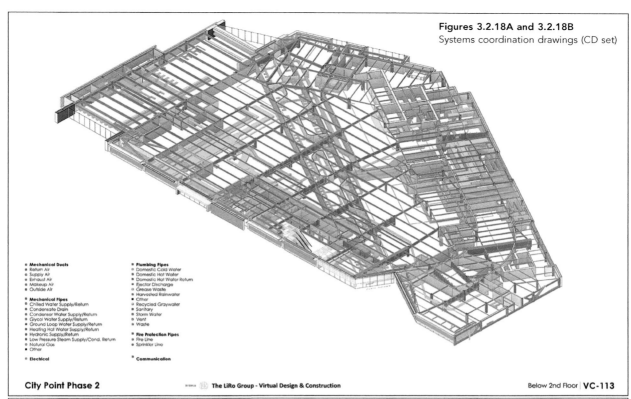

Figures 3.2.18A and 3.2.18B
Systems coordination drawings (CD set)

Mechanical Ducts
- Return Air
- Supply Air
- Exhaust Air
- Makeup Air
- Outside Air

Mechanical Pipes
- Chilled Water Supply/Return
- Condensate Drain
- Condenser Water Supply/Return
- Glycol Water Supply/Return
- Ground Loop Water Supply/Return
- Heating Hot Water Supply/Return
- Hydronic Supply/Return
- Low Pressure Steam Supply/Cond. Return
- Natural Gas
- Other

- Electrical

Plumbing Pipes
- Domestic Cold Water
- Domestic Hot Water
- Domestic Hot Water Return
- Ejector Discharge
- Grease Waste
- Harvested Rainwater
- Other
- Recycled Graywater
- Sanitary
- Storm Water
- Vent
- Waste

Fire Protection Pipes
- Fire Line
- Sprinkler Line

- Communication

City Point Phase 2   The LiRo Group - Virtual Design & Construction   Below 2nd Floor | VC-113

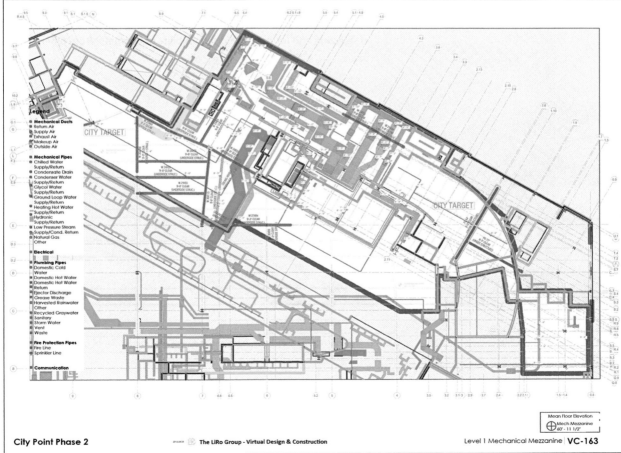

Legend

Mechanical Ducts
- Return Air
- Supply Air
- Exhaust Air
- Makeup Air
- Outside Air

Mechanical Pipes
- Chilled Water Supply/Return
- Condensate Drain
- Condenser Water Supply/Return
- Glycol Water Supply/Return
- Ground Loop Water Supply/Return
- Heating Hot Water Supply/Return
- Hydronic Supply/Return
- Low Pressure Steam Supply/Cond. Return
- Natural Gas
- Other

- Electrical

Plumbing Pipes
- Domestic Cold Water
- Domestic Hot Water
- Domestic Hot Water Return
- Ejector Discharge
- Grease Waste
- Harvested Rainwater
- Other
- Recycled Graywater
- Sanitary
- Storm Water
- Vent
- Waste

Fire Protection Pipes
- Fire Line
- Sprinkler Line

- Communication

CITY TARGET

CITY TARGET

Mean Floor Elevation
Mech Mezzanine
60' - 11 1/2"

City Point Phase 2   The LiRo Group - Virtual Design & Construction   Level 1 Mechanical Mezzanine | VC-163

Trade Coordination File Setup

Proper information model management is essential for a clear and organized coordination process. LiRo VDC uses the DWF file format for export from Revit to Navisworks. The default export format for transfer from Revit to Navisworks, NWC files, do not display in Navisworks with lines on the edges of the geometry. Consequently, a bundle of pipes or ducts of the same color is hard to differentiate. Another benefit of DWFs is that they can be updated while the Navisworks file is being worked on, while NWC files require the Navisworks file to be closed and then reopened to update. This also helped streamline coordination meetings, where having the most recent version of the file projected on the screen is essential. However, NWC does allow for grids and levels to be visualized in Navisworks, which is not the case with DWFs. Therefore a single NWC file is exported from Revit, so the grids and levels can be visualized in the coordination file.

Modeling of the individual trades was organized in Revit using its workset feature. Separate views for each Revit workset facilitate batch exporting of individual worksets as DWF files. These DWF files are then linked into the Navisworks Design Intent coordination model. It is beneficial to break each workset into a separate file in Navisworks for ease of selecting and for controlling the visibility of geometry from different trades.

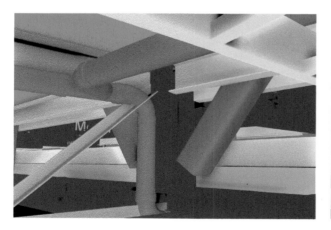
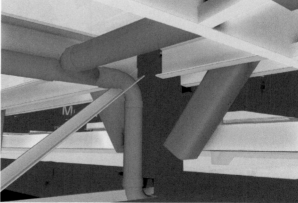

Figures 3.2.19A and 3.2.19B
Image of DWF vs. NWC file comparison. NWC and DWG files do not include lines, whereas they are included in DWF files (the preferred format, allowing better clarity)

Tips for Successful Coordination Meetings

A member of the VDC team should be available to drive the 3D and 4D models and facilitate communication about issues or clashes that have arisen in the project at all coordination meetings. During "live coordination," it is easy to get lost in the Coordination Model, especially when dealing with large projects. Time searching for an area under discussion or explaining to the team the location of what they are reviewing is wasted time. Using levels and grids as guides is helpful, as well as saving views for each room in the Coordination Model so any area can be accessed with a click.

It is recommended to use two displays during coordination meetings: one screen as support to show where the issue is in the building; and the other to focus on the specific issues in a perspective view. Plans are also useful tools for orientation.

Predefining visualization styles, or filters, greatly helps the coordination. For example, all the exterior architecture can be made transparent or opaque as needed. When there are ten highly paid people in a room it is the role of the VDC coordinator to quickly be able to locate areas of discussion, so there can be constructive conversation and quick resolution of issues.

Figure 3.2.20
Coordination meeting photo

Experience Simulation

As the design intent coordination of the podium progressed, the client approached LiRo VDC about creating a rendered experience simulation, or walk-through of the project. The walk-through was initially used to convey design changes and ideas during a client design review meeting. The client decided the animation would be useful for marketing purposes as well.

As LiRo had already developed the basic architectural model of the project, bringing the details of the model up to required levels for realistic visualization only called for a minimal amount of additional modeling. Lumion, a software built on an extremely fast game engine, generated a near-realistic walk-through animation, consisting of thousands of frames, in a fraction of the time it would have taken traditional animation packages such as 3-D max or Maya. To use the full power of the software, LiRo custom-built a computer with a GPU (graphical processing unit) optimized for the game engine. To create an animation of City Point without LiRo's architectural model, the developer would have needed to hire an outside visualization consultant to first build up the geometry, then render it, and essentially reproduce the modeling already done by LiRo. Traditionally, visualization consultants do not use BIM-compatible software—instead, 3-D models are constructed as mere geometrical representations and are not linked to the evolving information model.

Advertising displays and wayfinding signage were modeled in Revit into the information model. The graphics were provided as digital image files by the project's environmental design team, and applied to the Revit model as decals. The model was then exported to the visualization software where animated people, cars, vegetation, lighting, and other landscaping features were added. These elements provided a much more realistic experience when navigating through the spaces. Since the software has extensive libraries of animated people, cars, trees, and other features, bringing the project to life is relatively quick and painless. Furthermore, Lumion accommodates models of many file format types, making it possible to import models from various online 3-D model libraries such as Sketchup's 3-D Warehouse[1] and TurboSquid.[2]

The client's marketing department requested that walk-throughs be customized for each of the potential retailers. By creating custom depictions of the spaces, the marketing department was able to demonstrate layout alternatives and advertising opportunities for their prospective retailer clients as an experience simulation—a decidedly stronger sales presentation than the typical 2-D plans and pictures package.

With every update to the information model, it was possible to update the walk-through animation by re-exporting and reloading the Revit model. Thus, the animation became an ongoing tool for the developers to keep up to date with design changes, better understand revisions, and make better design-related decisions.

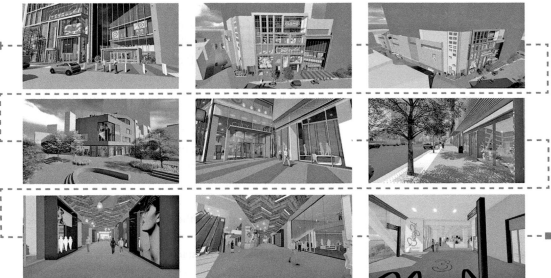

Figure 3.2.21
A series of images from the City Point walk-through, exported from Lumion

INTERVIEW WITH FRANK BONURA AND BRAYAN SILVA

Frank Bonura is a licensed architect and Vice President of Design and Brayan Silva is a BIM Manager for Albee Development LLC. This interview took place at the City Point field office on October 9, 2014.

F. B.: I manage the whole Architectural and Engineering part of the project. I negotiate all the contracts with the various AE consultants, and am involved from concept design to where we are today, with the construction administration phase. Brayan is my right-hand man as far as developing master plans, leasing plans and he's also gotten involved with the BIM coordination. During the design process, we sent him to training and that is how we heard about your company as we were interested in getting the modeling started on this project.

We recognized that due to the complexity of this project, with three different developers on-site, the coordination that needed to occur, the podium infrastructure and structural coordination, a 3-D model would be incredibly useful.

What was the biggest benefit of using VDC on City Point?

F. B.: I think there were a couple of things. One was that we initially used it for MEP coordination we were pleasantly surprised, or unpleasantly, depending how you see it, about the number of clashes we detected. I would have never imagined that we were going to have the number of conflicts that we had. So that was incredibly useful, to be able to resolve those issues early on. I think we engaged you when we were about 85 percent construction documents or so. By the 100 percent set, you had already caught up with the modeling and we were able to do a real coordination of the project in 3-D at the submission of the construction documents. We picked up things that we would not have been able to pick up from just looking at a set of 2-D drawings. One example that comes to mind is we were looking at the routing of a storm line in the concourse level running from Gold Street to Flatbush Street. And you just see a line going down the sheet, but when you factor in the pitch and the length, by the time you get to the other end it was just 5 feet above the floor, which is not enough clearance. Neither the architect nor the engineers thought of that, so that would not have been detected until you were probably halfway through the installation, unless the contractor possibly would have caught it.

The second thing is that we started to see all this progress, once we had modeled the structure, all the systems and interiors. We continued, let's do the facade, signage, and lighting and before you know it we had a real, virtually constructed model of our project. We were able go ahead and show these virtual tours to the partners [of Albee Development], and then make design changes for certain things that we felt didn't look right.

So it was useful as a coordination tool, and then later as a design review tool.

How would you have structured the project differently if you were starting over, knowing what you know now about VDC?

F. B.: To be honest, we did not use BIM out of necessity as the architect didn't have the ability to use BIM. If I ever do a large project like this, I will make sure to hire an architect that has the capabilities to model the project from day one. I think a lot of developers are in the same boat, they don't have a design team capable of BIM. The design team should be using this and they should be the ones that work through these issues.

In this case, the VDC team ran the coordination process. In your opinion, was this effective? Who should be responsible for VDC coordination?

F. B.: Someone has to coordinate it. Traditionally, it's the architect that does it and they should take the lead on this. An example on this project is Prince Street [the interior street running through the project]. We clearly stressed early on that it had to have a 20-foot clear to the ceiling. When we did the model, we realized that it was only 17 feet on one end. Where was everyone? Obviously the architect had not communicated with the mechanical engineer, who had just routed something on a piece of paper without understanding the ceiling height requirements. They clearly did not coordinate. If the area had been modeled by the design team, the limited ceiling would have been caught and resolved early on during the design phase. Ideally, at the end of the design process, if the AE team has done coordination, there should be no clashes when the construction documents are completed.

Looking back on the work that has been done, what could have worked better or been improved in the VDC coordination process?

B. S.: When we got to the construction phase, the subcontractors didn't have experience with BIM and didn't engage with coordinating the trades all in the same room. It probably would have helped to have much clearer contracts, with the BIM requirements for the construction phase written in the specifications, clearly outlining the responsibilities of the parties involved in the coordination process. Especially when some of the subs outsourced modeling to India, which caused a disjointed coordination effort at first.

F. B.: I think the fact that this project was open shop and non-union. The contractor is not as savvy. The tower two team were all union contractors and they seemed more on board with using BIM. Otherwise, I don't really see what the drawback is. It is a dollar issue: If you calculate what you are going to spend on the model and you minimize your number of issues, it does pay for itself many times over. One major structural issue alone can pay for the whole effort. We caught all the steel beam penetrations in advance and were able to prefabricate those off-site. It would have been a disaster if that hadn't been picked up.

> From a CM perspective, are there any tools like 4-D scheduling and BIM-driven cost estimating that you would like to use on your next project?

F. B.: I can't speak for the company, but on any other large projects like this one it would be greatly helpful to be able to see the schedule tied to the model and use it for estimating—to really take full advantage of all the capabilities. It has to prove itself though, I think people are hesitant to try and pay for something new if they don't know the outcome. I heard when it first came out, the big design firms were using it and they were having a lot of issues, but now people have perfected it.

> How would you quantify the value of the VDC coordination? It's hard to count savings on change orders that don't happen.

F. B.: This effort has more than paid for itself. Just after the first round of coordination, it more than paid for itself. That, plus the whole marketing aspect—not only did we do the design review, but then the fly-through that you produced we found was the best way to communicate the project overall. It helped people understand how pedestrians would navigate their way through it.

We also used the model to coordinate with the future tenants. It was useful to coordinate the clearances with the retailers, as those requirements were part of the standard fit-outs for retail chains. It enabled us to actually coordinate certain areas with even higher ceilings and really deliver what they were looking for.

> Now when things are being installed, are all the subcontractors actually installing their systems as per the signed-off coordinated model?

B. S.: The MEP superintendent is so into it now, so he actually measures the model and goes around verifying the installation. They had a problem with a standpipe, and he pulled up the model and could quickly verify that it was not installed as coordinated so they had to move it—no discussion. The sheet metal and sprinkler guys are actually manufacturing their installations from the model.

F. B.: It also helps with the as-built drawings. They will give us CAD drawings with all the elevations on it that the facilities managers can use, as they are not using Revit. The facade was also manufactured using BIM. The consultant engineered everything themselves.

Any additional conclusions about the use of VDC on City Point?

F. B.: Some of the clashes detected [in the VDC coordination] would have been extremely costly to fix. Considering the complex geometry of this project many of them would not have been caught until during construction resulting in costly, time-consuming change orders. So, yes, I think that overall the project would have taken longer and cost more if we hadn't implemented VDC.

CITY POINT PHASE II: CONCLUSION

City Point was unique in that it dealt with the modeling and management of the coordination process for three separate project teams within one concurrent site. The skills and active involvement of the trades are of utmost importance, in any coordination process, but especially for a project of such complexity. Any delay in the quality or timely delivery of any one trade's model affects the quality and delivery of the entire project, as it holds up the coordination process for everyone. For City Point, many of the subcontractors did not have experience with VDC, so there was a learning curve to get them up and running. This included acquiring computers and software capable of running VDC software, training staff, and educating decision makers on how to read 3-D models.

For an efficient, organized, and successful VDC coordination process, it is critical for the entire team to commit fully to the process. During the time the VDC team was involved at City Point, a wide range of serious issues were uncovered, which were solved by all the parties working collaboratively. One of the great values a VDC team can provide is identifying inconsistencies in the design that can be rectified before construction starts. The success of this process requires the participation of the entire team to come to mutually beneficial resolutions and follow up on updating project documentation. On this project, a range of changes took place because the client was able to understand the building prior to its construction through the use of experience simulations, and could provide valuable feedback during the coordination, resulting in a more successful building.

NOTES
1 https://3dwarehouse.sketchup.com
2 www.turbosquid.com

Figures 3.3.1A and 3.3.1B
Photos of the caverns after excavation

3.3 East Side Access

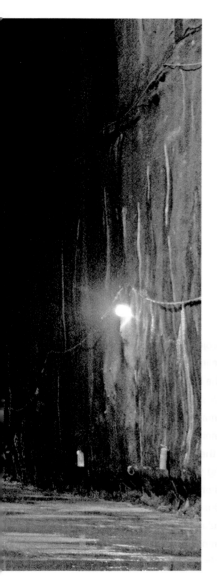

East Side Access is a new train station connected to Grand Central Station. It is a public-works mega-project commissioned by the Metropolitan Transportation Authority (MTA) in New York City. When complete, Long Island Rail Road trains will directly connect to the East Side subways and Metro North. The project includes the construction of new tunnels under Manhattan, a new eight-block-long concourse and eight train platforms beneath Grand Central Station at a depth of approximately 150 feet, the refurbishment of existing train tunnels under the East River and an above-ground train yard and interlocking in Queens. The new terminal is projected to be complete by 2022. The project is distinctive, as it is one of the largest and most complex infrastructure projects currently under construction in the United States. The VDC team client, MTA Capital Construction Division (MTACC), acts as an owner for the project.

Overall Project Budget:	~$10 billion
Location:	Manhattan and Queens, NY
Completion:	2022
Type:	Heavy Civil, Infrastructure, Railroad Station

Underground Construction

Client:	MTA Capital Construction
VDC Fee Structure:	Yearly budget approved at the beginning of each year
Contract Type:	Joint venture with Jacobs and LiRo Group
VDC Focus:	4-D scheduling; logistics and constructability coordination

VDC Technologies

Information Model Authoring:	Autodesk Revit 2014
Coordination:	Navisworks
4-D Scheduling:	Synchro
Visualization/Animation:	Lumion/Adobe Premiere
Post Production and Visualization:	Adobe Suite

PROJECT BACKGROUND

The LiRo VDC team joined the MTA project team more than ten years after the start of the project. In 2011, the team was invited to present for MTA's East Side Access Management after the MTA construction community became aware of LiRo's successful VDC implementation on another MTA underground project, the 7 line. The magnitude and complexity of East Side Access presented MTA management with numerous challenges in the planning and procurement stages of the project. Coordination of limited access points and passageways among multiple contractors during the active underground construction was one of the most pressing issues that needed to be addressed. The ESA program management team understood that information models and 4-D simulations could aid tremendously in the process of coordination and planning.

The ESA project is split into multiple contracts, which are staged to achieve smooth construction delivery. The VDC team was asked to deliver 4-D simulations to help with communication and in facilitating the planning stages of the procurement. Although LiRo's VDC team was hired specifically to create a 4-D model of the project, the team understood that the 4-D component was only a part of the larger VDC ecosystem that needed to be developed, with a well-organized and accurately created information model as its foundation.

Placing the VDC team alongside the program management team in the project controls office at ESA headquarters was a key factor in the success of the VDC effort. Being able to intimately engage with the organization made it possible for the team to design a VDC methodology tailored to the culture of ESA. This proximity also allowed the team to speed up the deployment stage of the VDC effort, and enabled VDC methodologies to grow and spread organically through the fabric of the organization. The embedded VDC team included two VDC managers and two VDC modelers. Additional VDC specialists were engaged at certain times to meet specific deadlines and deliver other additional VDC products.

The initial scope of the VDC team's involvement was to create an information model of the Manhattan portion of the project and to facilitate the creation of a 4-D master model depicting access to various underground sites. Due to the enormous size of the project, much of the team's initial effort went into developing a sufficient understanding of the

SUCCESSES
- Provided the client with communicative visual materials derived from the information model, enhancing operations and elevating discussion surrounding critical issues on the project, particularly in the areas of risk management and project planning. Specific issues were conveyed through customized diagrams and presentations.
- Developed a 4-D model for schedule visualization and inserted it into the standardized project control workflow.
- Connected all levels of operations with new technologies, enabling an effective flow of information to spread across the owner's multiple offices.
- Provided high-definition walk-throughs of the concourse and caverns, providing the client with a valuable understanding of future advertising potential and retail opportunities in the interior of the station.
- Developed a system to streamline data collection from the field operation to the tracking model for tracking of construction operations.
- Developed and integrated contract models and specifications to mandate the contractor's use of the information model as a tool to generate shop drawings and a 4-D model, with a final delivery of an as-built Facilities Management model.

CHALLENGES
- VDC methodologies were introduced when the projects were in the middle of the construction phase. This required a gradual integration of VDC methods into established workflow.
- It was difficult to get some of the stakeholders on board, as applying VDC on such a large project was uncharted territory.

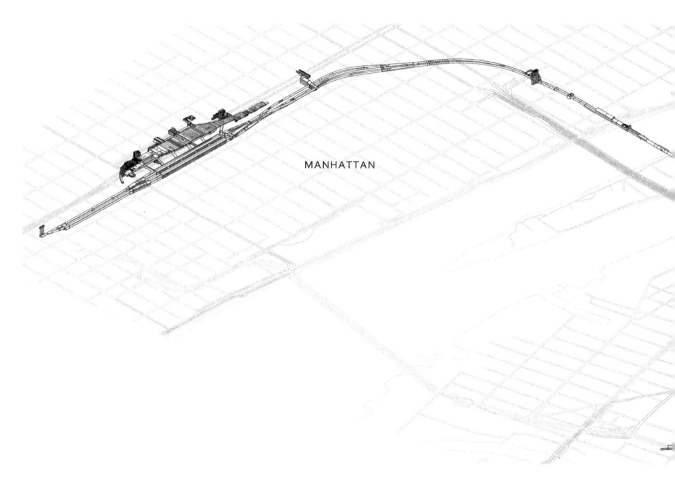

MANHATTAN

Figure 3.3.2
Diagram conveying the extent of the East Side Access project: 6 miles in length, ESA includes the construction of a new transportation terminal spanning eight city blocks, multiple ventilation buildings, 7.8 miles of tunnels, and a large rail interlocking in Queens

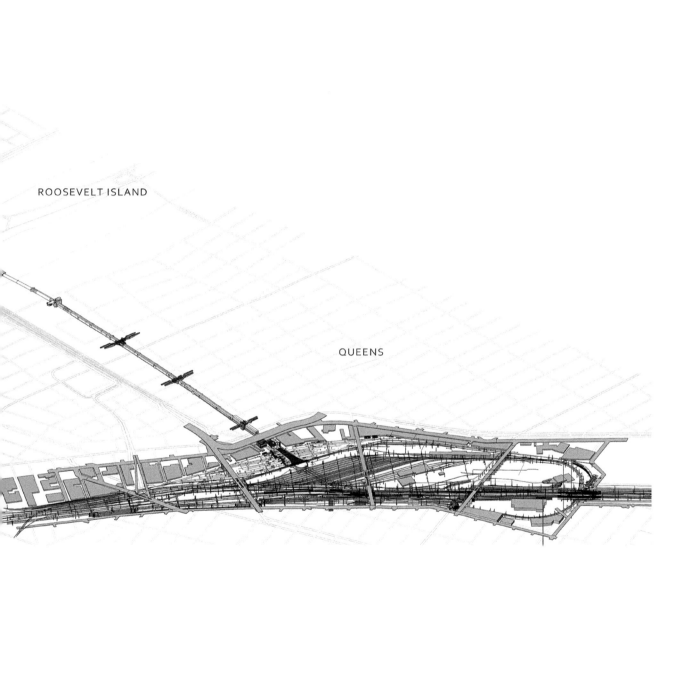

ROOSEVELT ISLAND

QUEENS

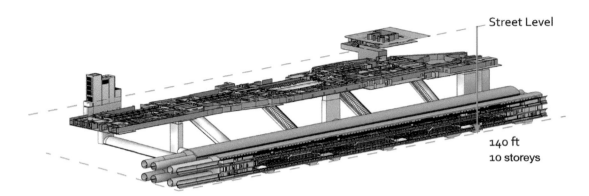

Figure 3.3.3
Side view of concourse and caverns,
showing the depth below street level

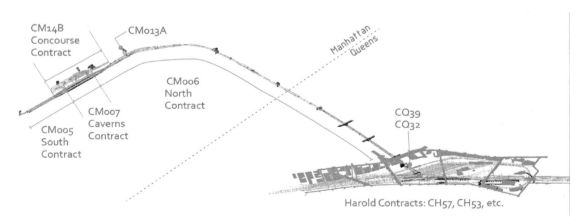

Figure 3.3.4
ESA has been broken up into multiple
contracts, because of its size. This
diagram specifies the contract areas of
ESA for which VDC has been used

ESA Organizational Chart

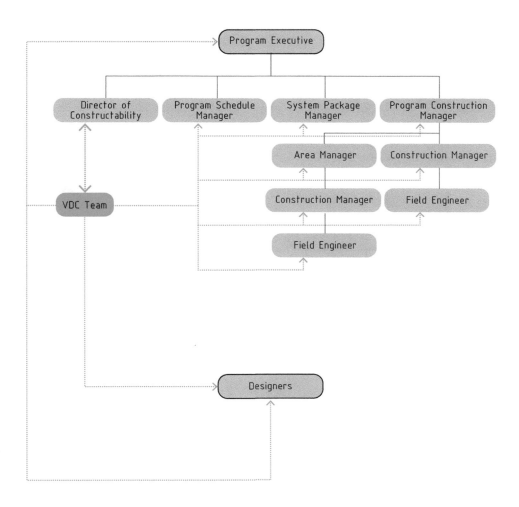

Figure 3.3.5
Diagram showing project team organization

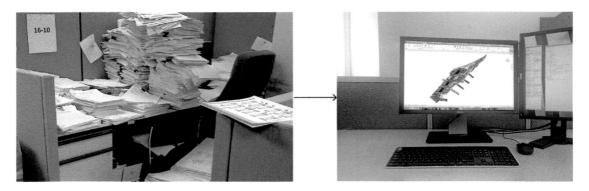

Figure 3.3.6
2-D analog paper process versus 3-D VDC process: All information including plans, drawings, tracking data, and schedules are incorporated into the information model

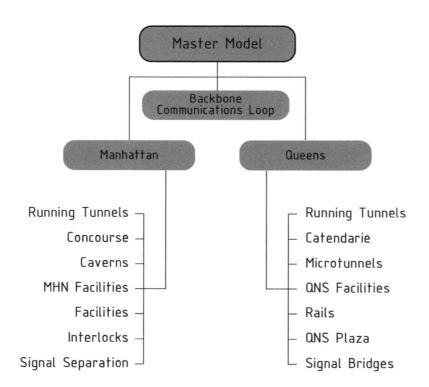

Figure 3.3.7
Organizational diagram of the variety of models that make up ESA

complex nature of the project, in order to begin modeling effectively. Over the course of the first year and a half of engagement, the VDC team transitioned project information from 2-D to 3-D format (Figure 3.3.6). The project was analyzed by the VDC team and broken into almost 50 Revit models, each of which represented a single facility. These models were linked by shared coordinates into an ESA master model, which allowed the models to be studied in context (Figure 3.3.7).

To introduce the owner to VDC and explain the modeling process, the VDC team created a series of posters depicting structures and areas of the project as they were completed. These posters were instrumental in orienting the owner to the 3-D model, and in developing a 3-D language for understanding a project that had previously only been visualized in 2-D. The posters were highly effective for impromptu meetings and discussions surrounding individual facilities. ESA managers and engineers working on the structures had never experienced their project in the 3-D format, so it provided a new perspective for the project teams. These posters migrated into the conference rooms and throughout the owner's job sites, becoming the central visual aid used in supporting technical discussions.

Figure 3.3.8
Posters in the client's conference room

Figure 3.3.9
Images from walk-through animation. High-definition walk-throughs of the 48th street entrance, concourse, and caverns provided the client with a valuable understanding of future advertising potential and retail opportunities in the interior of the station

After the initial success of the structure model, a systems model was initiated and developed to visualize the major equipment and access routes for equipment installation throughout the entirety of the program. The VDC team created a 4-D project-wide model and implemented a project-wide VDC ecosystem. The VDC scope also evolved to include support of multiple construction management field operations, as well as the development of contract models for construction operations.

Since a comprehensive 3-D model of the East Side Access project already existed as a product of the 4-D effort, the client requested a walk-through animation of some of the Manhattan structures. Additional detail, signage, materials, and faux retail spaces were added to the model to boost the realism of the experience simulation. The resulting animation served a two-fold purpose. Initially, it allowed project architects and management to accurately visualize, in high definition, all the spatial components and finishes together. Second, while collaborating with third-party advertising consultants, the animation served to illustrate the best scheme for integrating profit-generating dynamic advertising displays.

INTERVIEW WITH MARK DEBERNARDO

Mark Debernardo is the Director of Constructability and Planning for MTA Capital Construction. Mr. Debernardo was interviewed by Lennart Andersson in the East Side Access main office on September 24, 2014.

What is your role in the project?

M. D.: My official role or title on this project is Director of Constructability and Planning. Being able to understand and recognize the risks and challenges to any sort of build-out is a critically important skill. As is conveying those challenges to the people who may not be as technically oriented or understanding of the skills it takes to control, establish, and track a project.

I am basically on this project going all over the place trying to understand what problems people are having with the build-out and the schedule that we have had over the last 16 years, and trying to keep them current and ready to tackle the challenges of the build-out as it's going today, and how we see it going in the years down the line. This is where my function lies. I try to help people understand the vision and the direction, and try to help them make management decisions in a timely fashion. It is a huge machine with many, many moving parts. Lots of people in the audience that you have to appease to move forward.

Did you have any previous experience with BIM?

M. D.: No, and I am not shy to say that, either. It is interesting how it was presented to me. I had read many articles about it and understood that it could be something that could be integrated in the future with CPM schedules, in which I have an extensive background. But, I had actually never been in an environment where it was utilized.

Why was BIM brought into this project?

M. D.: This is a very large mega-capital construction project. It is the largest transportation project I know of in North America. It's very complicated in terms of the participants that are needed to execute the build-out, starting with all our railroad brethren, who have territorial rights to this project. We go into their work zone. We have to disturb and manipulate their revenue service operation, which cannot be taken out of service while we are building this new arm to their route system. Amtrak and Long Island Rail Road are huge power entities in commuter rail and freight. Getting them to understand and stay in sync with what we need from them to facilitate a build-out in a timely fashion within budget and a schedule, we started to realize we need better ways to convey the message and get decisions out of people like this. In all honesty, we have been spending a lot of time drawing hand graphics in order to get concepts across to people. Railroad people that run train services are not necessarily engineers or technical people who really have the background to read technical two-dimensional drawings.

When was it decided to use BIM on this particular project?

M. D.: There was an effort to start using it four years ago, which went on for about a year and a half. I was aware of that effort being made, but I was not directly involved. I was brought in with another group of individuals; we were looking at a pop-up product coming out of development that was perceived to be a building information model. At the time, the scope of work and our true understanding of the vision was not clear. I think that led to an unsuccessful sort of exercise, where we didn't reap the full advantages of the product, because it did not address what we thought it would. For the most part, we were trying to use BIM to tell people how we would move material between Queens and Manhattan while progressing the build-out of the tunnels themselves. We have multiple contractors working in a very confined environment. We had to prove to our stakeholders how we were going to control and manage that. Our first effort with BIM didn't really accomplish that as well as we had hoped. We took a hiatus, and eventually terminated the effort. From the beginning, I thought the way it was kicked off was not really as well thought out as it should have been. So I just marched on in terms of how I expressed my work to other people; I used other mediums that we had been using forever in the industry—schedules, estimates, simple graphics, two-dimensional drawings marked up with color highlighters—trying to make people understand what they are looking at.

What ended up happening was we walked away from it (the first BIM effort, which was not successful), and a couple months later the current VDC team was brought in to start a new effort. I think there was a recognition that this team had people that were very skilled in VDC technologies, and our project executive felt that we had to give this another shot, as we were still having the same problems that we had before. Things were getting more complicated with a lot of construction already underway, and there were a lot of places where we needed people to understand what was going on. It was very difficult to do and to keep up with the timing of the decisions that needed to be made.

The new team was brought on board and my boss at the time came into my office and told me, "Guess what? You are now the new BIM manager at East Side Access." My first question was, 'What does BIM stand for?' [laughs] And that is how it started.

Given the little knowledge you had about BIM, how have your perceptions changed?

M. D.: Tremendously. When I started, I realized this field is quite unproven—only something you had read about in publications and magazines. Something a lot of marketing people talk about. It was rare that I met anybody who actually did it. I am a long-time New Yorker and my niche in the industry is mainly in transportation, where we haven't seen a lot of this. If you go into the building industry, especially in the private sector, you probably encounter a whole lot more of this technology. However, I think the advantage I had was that I recognized the need for superior graphical representation of a project. Something that has the capacity to be manipulated without losing its true properties or dimension, and that can help people understand what the talking issues are and see what we were asking them to understand from contract or engineering drawings. So we put a little plan together—the deal was for the VDC team to teach me what this is all about what they have been doing all their lives, and I am going to get them through this maze of a project. We are going to start with addressing the immediate needs, and as we do, we can also plan out how to be as effective as possible moving forward.

What do you see as the potential of VDC?

M. D.: I think now, a little more than two years from when I started, I would not call myself a VDB expert. I am not someone who can go into the software and actually create a model. I have worked with a host of people, I have studied their mannerisms and skills and how they use VDC to convey messages to me. That was the biggest challenge. If they could sell the ideas to me, I could turn around and sell it to anybody else. That was really key, as I would not take the stage unless I was confident in what I was going to show. I think the whole industry is really far behind the ball. We experimented with many things and took risks with how we dove into many different corners. I can't think of a single challenge where we did not reap a quality benefit from it.

I will tell you in most cases we exceeded people's expectations as to the benefits they would witness coming from it. We have gone into claims, construction methodology and animation; we have created a new level of reference material using VDC in the very simplest of fashions, with still images. We created whole books of images and organized them such that people could walk through the alignment in a very methodical fashion, to see what the structures looked like, and look at the different sections. If there was an area they could not see, I could show it to them with a live model. Then we created an easier way for people to interact with the model, by creating models in Navisworks, which people could interact with from their desktops. We set up simple instructional classes, which were never more than an hour or two. After that I had people calling me up telling me, "This is great!" People have the model on their desktop, and there's no way they can mess it up, but they can maneuver and cut it up and create their own images based on what they need to look at. We got involved with supporting large procurements on this project; through our process we've been able to give models to the contractors, which they can open up to see the structures. They really appreciated this, as it gave them a much more proportional view of the scope of work and helped them assemble their estimates. In the last two years, we've seen a significant trend toward better estimates that are closer to the expected budgets we have established for these jobs.

We didn't stop there, we tried to challenge ourselves. We asked ourselves, what is the next level of this? I always had in the back of my mind a retroactive approach to applying this on East Side Access. We started to push the notion of integrating the CPM schedule with the information model for 4-D scheduling and tracking. We started out on a simple concrete job. We built a good rapport with the contractor, and they saw the benefit of visualizing the job's progress week by week. We could actually reflect in the model where the job stands. It is tied to the CPM and now we actually have it tied to the payment schedule. I am probably most proud of that, because that, to me, is a really classic application of BIM. It's not that it's difficult, it's like anything else: You have to plan upfront, and if you think about how to utilize and integrate it into scheduling and cost estimation it works. It actually really does work. We have now gone on to the next job with the lessons we've learned and are continuing the process, so it has really started to take off.

The 3-D graphics are unmatched—it's the wallpaper of this entire project at this point. We didn't stop there, either: we are now going in the direction of integrating VDC specifications into contract documents, forcing contractors to use VDC to manage their cross-disciplinary coordination. We will be able to see and address through shop drawing submittals our first challenge, which is always whether it is dimensionally correct. Does it fit? Is it coordinated with all the other entities involved in a ceiling space such as the HVAC, electrical, plumbing, and fire protection? We are about to launch the first specification.

We are also going through the process of creating an existing conditions model. If you asked me three months ago, I never thought we would have been doing this. It scared me to death a bit, just to say that we are taking this on. We hired a specialist that does 3-D scans and converts them into a point cloud of the existing conditions. He gives it to us, then the VDC team creates a Revit model from it. Once created, we take this existing conditions model and plug it into the design intent model based on the 2-D drawings. That's when we see if there are any bumps or hits, called clashes, which is always the scary part. But it's a great quality assurance test to see if everything fits. And if it does not fit, you are better off knowing now than finding out later out in the field.

We seem to have very good momentum right now, be heading in the right direction. Every day that goes by, I feel more confident that we are doing the right thing. We convinced our CMs this is the right thing to do, and we are trying to help them understand how we are going to manage this when we get the contractor on board. Then we will all go through the process of what we're actually going to do. This to me, again, is another classical application of this technology.

We probably spent the first two years with a sort of ad hoc attentiveness just to emergency needs, and it's really been interesting how we've since stepped up to the table. Again, I still look at this as something that needs to develop in concert with all the other construction and engineering skill sets needed to control jobs like this. Throughout the design process, throughout the procurement and construction process. And then through the claims and mitigation and risk management frontier. I am a risk manager here, too, and I can't tell you how many times I brought the model into the room to discuss and reinforce issues and then to show alternate scenarios for how we might manipulate and/or alter something. Again the power of VDC, with the proper type of planning and integration with the team, is having the information readily available, if you know how to structure it. It is all about pulling the team together and having the right type of squad. I am very proud of that and where I see it going.

So looking at the other side, is there anything that VDC does not do or problems that arise when using VDC?

M. D.: There are. I'll tell you the number one negative is there are not enough people using it. There are not enough resources that understand it just yet. A lot of people come to me, I've built a bit of a notoriety now: Mark is the guy to go to. I have other projects calling me asking, "How did you approach this? How did you find the people?" I have people sending me their resumés. They are telling me that they know Revit and that they are BIM specialists, but their resumés and portfolios don't show it. I'm getting a lot of graphic artists.

It's interesting, I was the guy who didn't know what BIM stood for a couple years ago, and now I'm interviewing people and telling them they have a total misunderstanding as to what BIM is. It's an arm of the engineering and construction community that needs to continue to grow its muscle. It needs more of its user groups and community established so it can be better publicized, shown, and demonstrated that this plays a real role today. I could not visualize this project without BIM—I don't know how we did it so many years without it.

If you started the project today, what would you have done differently?

M. D.: I am a true believer in planning, so if we started with a conceptual plan or layout strategy, I would consider BIM as part of the design process in working with the designers. Not only do I want to see a two-dimensional drawing set, I want the information model, too. I want to use it for constructability reviews and major design submittals, which are reinforced again with models. BIM is a valuable asset at the beginning of a project. Especially for jobs of magnitude, where you are trying to fight for funding and describe how this job looks. You're going to the people that have the money, but they don't necessarily have the background to understand how exactly things are done. Having a model like this can help them visualize. We did this in the blind for many years here at East Side Access—close to 10 to 14 years. It's amazing.

Do you think if you had these tools that help understanding from day one, the project would it be further ahead?

M. D.: In a classical application, I always have the main model that I can go to and copy, so I can start to do my derivative if/then scenarios. When we chose, almost a year ago, to defend two very significant claims here at ESA, we did not have the model developed enough. We were not really ready to start these initiatives, but we told everybody in the VDC team to stop what they were doing, we needed to put in a concentrated effort to generate the model and replicate the conditions that were needed to defend the claim. We had two shots at this, and were very successful both times. I was part of the Dispute Review Board when we were trying to convey our position on a very difficult claim, and the only people in the room that could speak and reinforce their concepts were those who used the model and showed it to the lawyers. Otherwise, no one would have been able to follow the conversation. Without you living the job and without using the model, you could not possibly understand the conditions and issues at hand. We used the model in such a way that we could turn on and off the what-if scenarios, and we very effectively got our point across as to where the issues and responsibilities lay from our point of view.

The biggest monster we have created here is the demand. At first, no one paid us much attention, aside from some of the simple things we did, such as the image reference documents. But then we sort of caught their attention. It's like when something on YouTube goes viral and everyone starts clicking on it. All of a sudden, everybody wanted us and we got stretched in all sorts of directions. That was a huge issue; that's why I say the biggest problem with the industry right now is we need to further cultivate people who are ready to enter this field so we can underpin. It is a huge tool that ties into everything. We are looking four to five years down the line on asset management and preparing information models for that. A year and a half ago, it was not a thought in anybody's mind.

With everything I have on my plate, I must admit that when I go over to the BIM group I feel like I walk into a different corner of the ship here. I can go in there thinking of a potential problem, and I can convey this problem. I have the tools, buried somewhere within the effort we have developed with the team, to replicate the problem and demonstrate it to everyone. That's refreshing, because it's very hard to do that sometimes on a project like this. It's been very effective.

VDC PRODUCTS

Design Intent Model LOD 300 level model that includes architecture and structure as depicted in the construction documents.

Realistic Walk-through of Concourse and Caverns The information model was used to create a realistic simulation of the station concourse and platforms, with a focus on the location of advertising, murals, signage, and storefronts. This photorealistic walk-through was the first time the project designers experienced the project, and it was used for a variety of purposes from marketing and community outreach to internal owner communication.

Systems Visualization The information model was used for feasibility studies to examine wiring and signal configuration, as well as spatial simulation to understand required clearances for people and trains in the tunnel. As the project progressed, the model was expanded to include all of the equipment in the major facilities. Due to the spatial constraints inherent in underground construction, the access routes for the installation of the equipment were modeled along with the pieces of equipment themselves. Extensive visual material was developed, then compiled and organized into book format. This became an important document for the entire project team's understanding of the systems. During the procurement process, the owner distributed the VDC system book to potential bidders as well.

High-definition Contract Model The existing conditions of the former underground rail yard that will house the ESA concourse needed to be documented in higher detail. MTA hired a surveying company to provide 3-D scans of the eight city block area. The resulting point cloud was then used to create a highly accurate existing condition information model. Architectural, systems and structural design intent models were brought to LOD 300 standard and provided to contractors at the award of a contract.

VDC SERVICES

4-D Simulation of Major Contracts ESA's major construction contracts actively use material generated from 4-D models to organize construction operations and communicate weekly construction progress. 4-D models allow schedules to become live, working documents, thus reducing the time it takes for the project control team to understand the schedule. This results in a collaborative environment in which decisions are transparent and easy to understand.

Risk Management Support An additional application of the 4-D model is supporting risk workshops, during which the owner's upper management reviews the overall project schedule and assumptions for the contracts. In these meetings, the 4-D model is operated "live" by a member of the VDC team, and has proved instrumental in bettering everyone's understanding of complex issues that arise during the project. Workshop participants can immediately see and agree upon problems that need to be discussed and focus on understanding the solution. This has been particularly beneficial to high-level managers, who are less familiar with areas of the program that are outside of their designated contracts.

Logistics and Constructability Studies The VDC team used the information model to create detailed studies of specific areas of the project, each designed to increase understanding of the complex issues associated with construction sequencing.

Litigation Support The information model has also been used in the support of construction claims. The 3-D models give an overview of the area and facilities in dispute, and 4-D models demonstrating the discrepancies between planned and actual work carried out on-site were invaluable in helping the owner make their case during mediation. Furthermore, the 3-D model was used to bring the legal team up to speed on key issues clearly and efficiently.

Virtual Prototyping: Cavern Architectural Panel Study The walls of the underground station (the caverns) comprised 432 precast concrete panels. The original geometry of the panels included two doubly curved surfaces, which were difficult to visualize using 2-D documents. The VDC team created a 3-D model of the cavern, which enabled the owner to understand the geometrical configuration of the panels. In addition, the VDC team created a 3-D printed physical model of the panel for the purpose of conveying panel geometry. Due to budgetary and logistical constraints, the owner decided to redesign the configuration and shape of the panels. The 3-D model was then used as an instrument for redesigning the panels, enabling the design team to explore multiple configurations to determine the most appropriate solution, as well as find the best-suited rhythm for several types of panels to fit together.

Construction Tracking The VDC team developed a methodology to track progress and visualize construction operations using a specialized construction tracking information model. This information model was updated daily to depict the most up-to-date state of the construction site.

PHASE | INFORMATION MODEL | VDC SERVICES | VDC PRODUCTS

CONCEPTUALIZATION

DESIGN

PROCUREMENT

PRE-CONSTRUCTION

Design Intent Model

Construction Management + VDC Models

Construction Models

Design Coordination

Experience Simulation

Project Logistics & Constructability

Distribution models

Systems visualization set

Walk-throughs/ Fly-throughs

4-D Simulations

Virtual prototypes

3-D Printed models

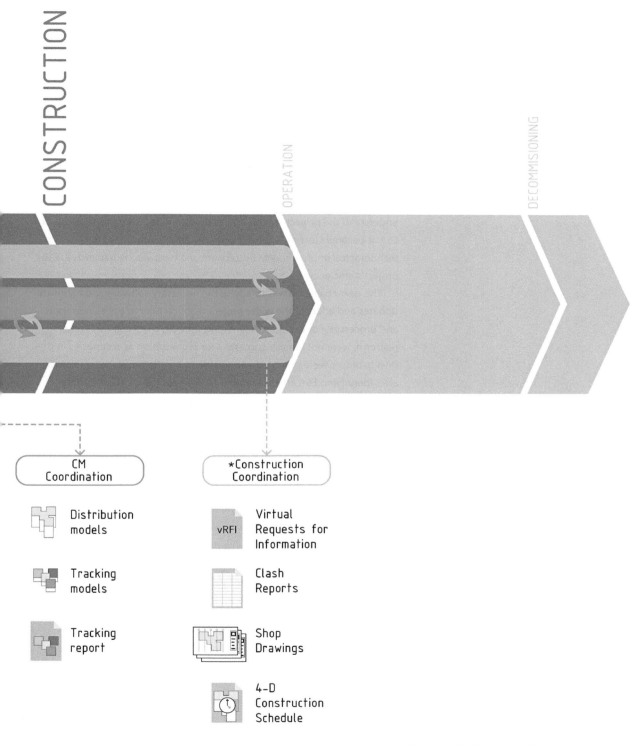

CONSTRUCTION

OPERATION

DECOMMISIONING

CM
Coordination

*Construction
Coordination

Distribution
models

Virtual
Requests for
Information

vRFI

Tracking
models

Clash
Reports

Tracking
report

Shop
Drawings

4-D
Construction
Schedule

* Construction coordination by contractor

VDC STRATEGY AND DELIVERY

The development of a VDC strategy takes time and effort. It is challenging to develop a sound strategy and robust digital infrastructure while simultaneously providing value and proving success. The key components to developing a VDC strategy for ESA were understanding the existing culture and workflow, amending it to incorporate VDC technology, and educating the team on the technology's capabilities, all while modeling the project.

Colocation in the client's offices is an effective way to understand their existing culture. This was absolutely necessary for ESA, because the project's size and intricacies required complete immersion in order to fully understand the project. As the VDC team developed an understanding of ESA, it became clear that better communication platforms as well as technological improvements in software and hardware would increase the project's efficiency.

The next challenge was to advise the necessary software and hardware updates and educate the wider project team on how to utilize new tools and processes. For example, Navisworks models, a key communication platform, were not initially successful for organization as computers in the on-site offices were outdated and unable to handle the software. However, after identifying ESA employees that would gain the most from these models, upgrading their hardware, and training them in how to use Navisworks, the models were extremely effective and communication of project objectives greatly improved.

Another major component of the VDC strategy was the mammoth task of translating thousands of 2-D drawings and verbal instructions into a 3-D Revit model, and with it, creating a visual lexicon by which the entire project could be understood (Figure 3.3.12). All VDC products were delivered as static images in PowerPoint presentations, a familiar and widely accessible format for the client. Animations, such as walk-throughs and 4-D sequencing, were also embedded into PowerPoint or delivered separately as mp4s. The Naviswork model was presented "live" in meetings by VDC team members, or delivered to project members who had access to the Navisworks software (Figure 3.3.13).

Figure 3.3.10 (previous page) Diagram of ESA products and services

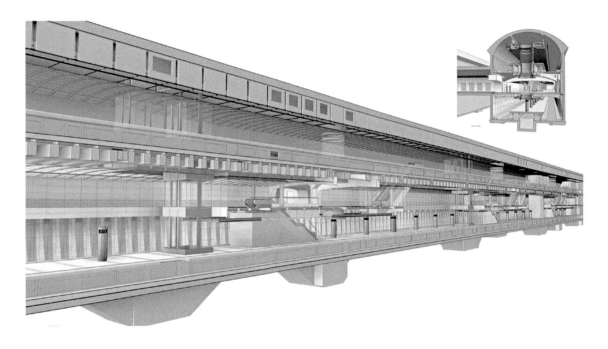

Figure 3.3.11
Image from design intent model

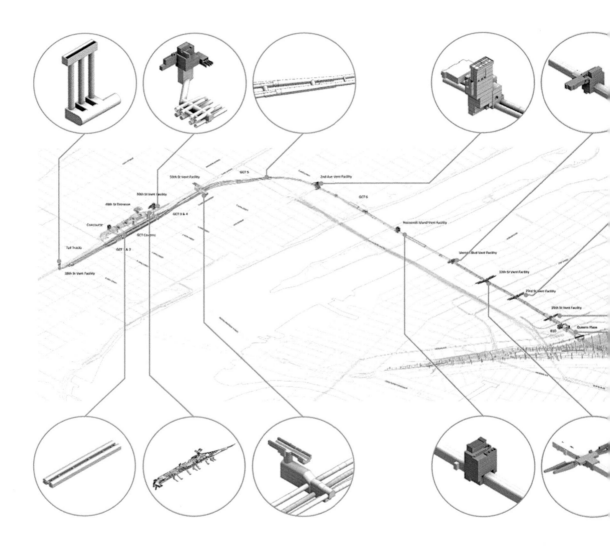

Figure 3.3.12
Images of facilities exported from the Revit information model. It was the intention of the VDC team that all ESA structures and facilities be viewed from the same angle or orientation point. These images became icons—a visual lexicon by which the entire project could be understood quickly and comprehensively, both when looking at the facilities individually, and as parts of the ESA

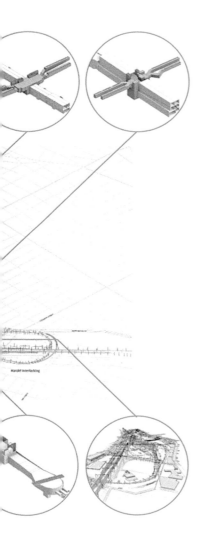

Use of 3-D and 4-D models extended beyond understanding the geometry of the project. They also helped in conveying various logistical issues that arose, and in critiquing alternative schedules and construction sequences. As such, the models were integral for risk workshops and litigation purposes; the team received requests for models in support of all decision-oriented meetings. Additionally, as the MTA grew to trust the VDC team and the accuracy of the 3-D data, the organization gradually became more open to innovation and experimentation with various technologies. Over time, the VDC ecosystem extended to include participants outside the immediate organization, including contractors, subcontractors, and outside consultants. The success of VDC on a project is defined by the ability of the VDC ecosystem to integrate and assist different parties involved in the process of construction.

Choosing 4-D Software

Prior to committing to Synchro for the project's 4-D deliverables, the VDC team researched the market for many possible solutions. There are many factors to consider when evaluating a new software for project use. Some of the major considerations are the standard and specialized software features, ease of interaction with the software's interface, maturity of the software, level of software support offered, the software's ability to integrate with other software already in use on the project, features that may be included in future releases, and the existence of a robust community of users.

Autodesk Navisworks was a possible alternative to Synchro and would have integrated well with the Revit model, as they are both Autodesk products. However, after testing Navisworks, the VDC team found that the rigidity of its interface does not allow for smooth interaction with and flexible set-up and control of the model. Furthermore, Synchro was far superior at generating high-quality graphical material. Synchro's intuitive interface and flexibility as well as its ability to handle large models were the deciding factors in the VDC team's choice to use Synchro as the 4-D solution for the project.

Figure 3.3.13
4-D model being operated "live" by VDC team member

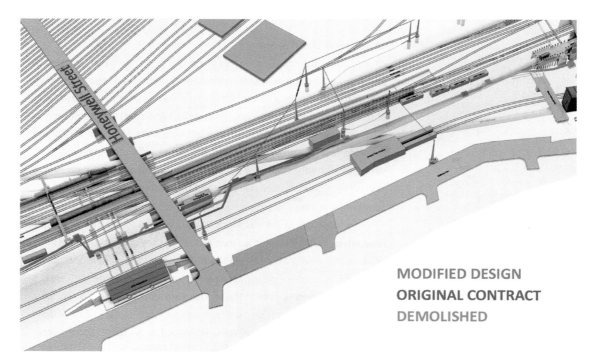

MODIFIED DESIGN
ORIGINAL CONTRACT
DEMOLISHED

Figure 3.3.14
The model in support of a claim

Manhattan

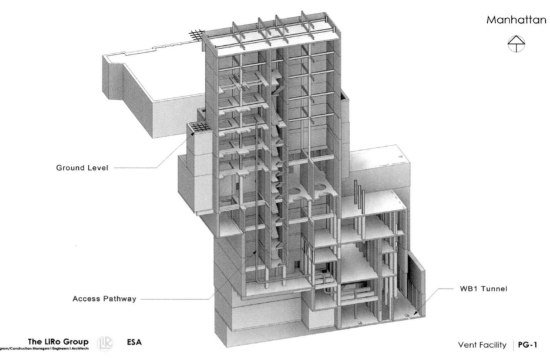

Ground Level

Access Pathway

WB1 Tunnel

The LiRo Group ESA
Program/Construction Managers I Engineers I Architects

Vent Facility | **PG-1**

Figure 3.3.15
An example of an image created from
the information model for a risk
workshop. The image is one of the
ESA vent facilities, highlighting a
proposed access route for mechanical
equipment

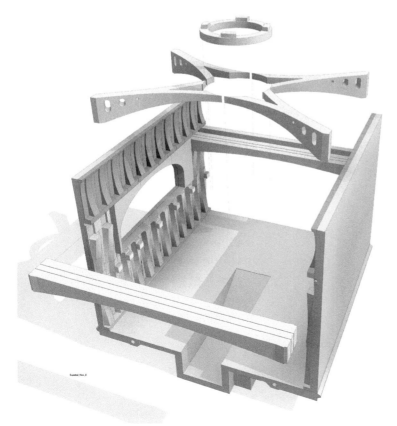

Figure 3.3.16
Exploded axonometric image
illustrating how the prefabricated
panels come together

ESA Software Tools

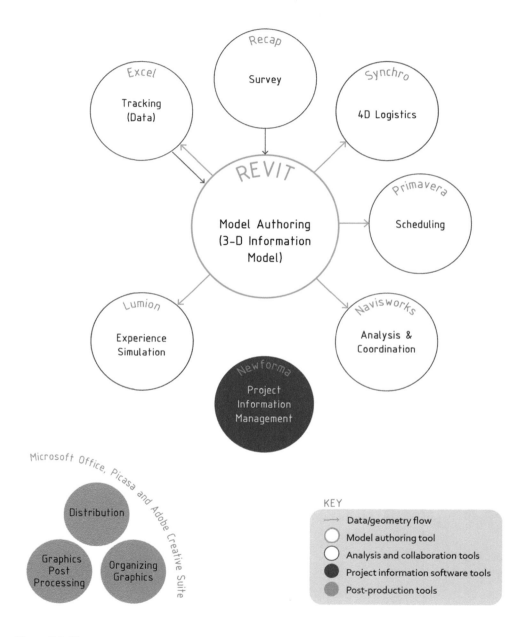

Figure 3.3.17
Diagram of software tools the VDC team used on the project

IN-DEPTH DESCRIPTION OF SELECTED VDC PRODUCTS

Harold Interlocking Model

As the VDC team became more integrated with the owner and more knowledgeable about the structure of the organization and the project, additional models were developed outside the original modeling scope. One such model was of the Harold Interlocking, the largest train interlocking in North America, which spans more than a mile and includes hundreds of miles of track.

An interlocking does not lend itself well to modeling as a traditional architectural information model, as it is a landscape containing buildings, roads, train tracks, tunnels, horizontal structures, many catenary poles, and above-ground and below-ground utilities. It took the VDC team six months to complete the first iteration of the 3-D model of the interlocking. As the first version of the model came to life, the owner was able to better understand the complexities of construction in the interlocking, which had been a very difficult part of the project.

The model was further developed as needed to support the coordination of complex logistical operations involving Amtrak, Long Island Railroad, and multiple contractors. It was also used to visualize and

Figure 3.3.18
A rendering from the Harold Interlocking information model

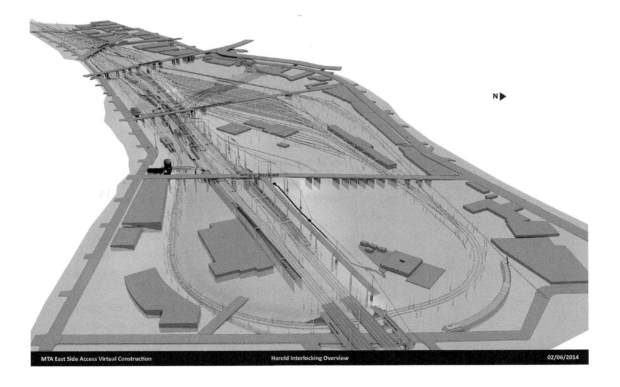

MTA East Side Access Virtual Construction   Harold Interlocking Overview   02/06/2014

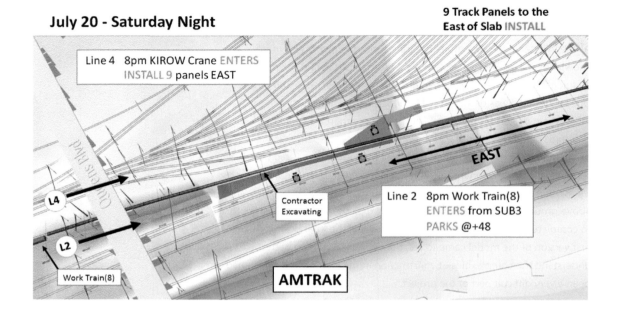

Figure 3.3.19
Slide from description of track
construction outage

explain the sequence of equipment and construction operations during
key timed track outages. The VDC team's agility and the structure of the
model allowed them to respond quickly, even to unforeseen uses of this
unusual type of information model.

4-D Simulation

A 4-D model illustrates a sequence of work by linking the fourth
dimension, time, to 3-D data from the information model. Geometrical
elements authored in a modeling software are linked to schedule activities
generated by a scheduling software. A third software is used to connect
the two types of data. In the case of ESA, Revit was used to author the
3-D content and Primavera was used for the schedule. Both the schedule
and the geometrical data were imported into the 4-D platform,
Synchro.

Synchro is used to generate three types of output—animations,
images, and an interactive "live" model that requires a special Synchro
viewer. More technical information about Synchro Platform can be found
in the Templates and Workflow section of this book.

Effective collaboration is a key aspect of a successful VDC process.
The 4-D model promotes collaboration by enabling the project team to
understand the schedule more clearly. It simplifies the analysis of complex
operation sequences containing hundreds of activities. Integrating

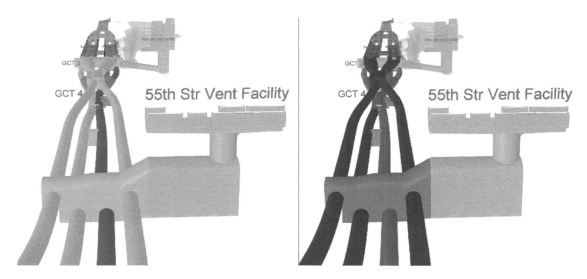

Figure 3.3.20
Images from 4-D model. 4-D output was initially distributed through sequences
of static images

multiple 4-D models together made it possible to achieve a centralized
4-D ecosystem, where the decisions made on a single contract are
available to project teams on adjacent contracts through the
synchronization of 4-D models. VDC field coordinators colocated at the site
office were skilled at operating the 4-D model and integrating it into the
CM's daily operations.

Construction Tracking

The first iteration of tracking methodology in ESA was developed to
facilitate the monitoring of construction operations of a power substation
in Queens Plaza. A major facility located in the New York Borough of
Queens on the border of Harold Interlocking, Queens Plaza is the last
location where the trains can switch tracks before they begin their ascent
above ground.

The VDC team assisted the CM with the visualization of construction
tasks in weekly increments. The main challenge in tracking construction on
the substation was conveying the complex geometrical configuration of its
structure. The model was positioned to be a central element in the CM's
tracking workflow. Images were generated weekly from the information
model to supplement the CM's report.

At Queens Plaza, the tracking effort was focused on the steel erection
operation. Daily tracking inspector sheets were supplemented with an

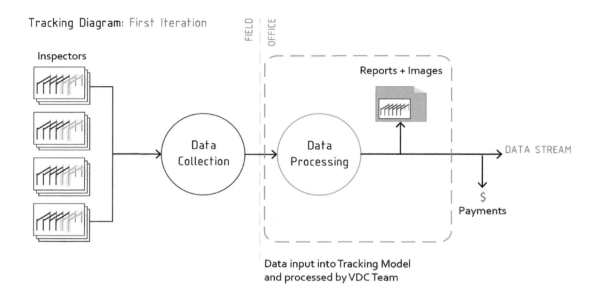

Data input into Tracking Model
and processed by VDC Team

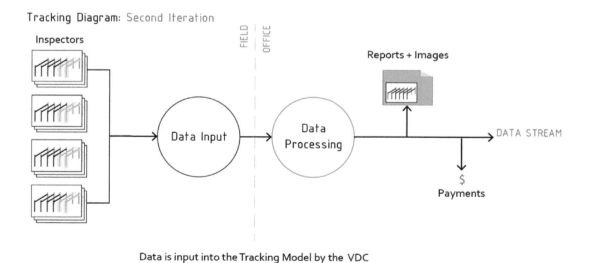

Data is input into the Tracking Model by the VDC
field coordinator and processed by the VDC Team

Figures 3.3.21A and 3.3.21B (i)
Diagrams of data collection and processing for tracking

Tracking Diagram: Third Iteration

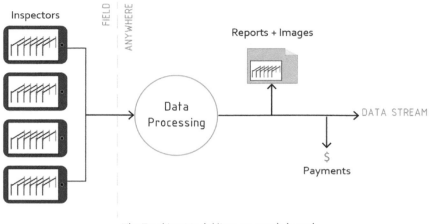

The Tracking Model lives on a web-based
platform, and is automatically updated by
inspectors via parameters set by VDC Team

Figure 3.3.21B (ii)

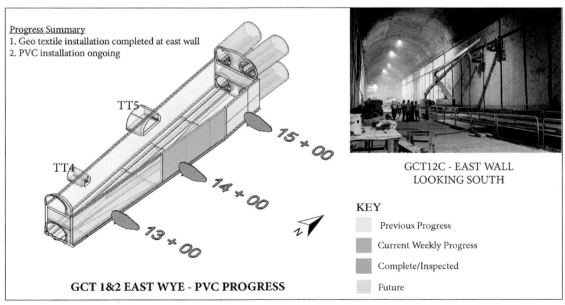

Figure 3.3.22
Example of construction tracking report

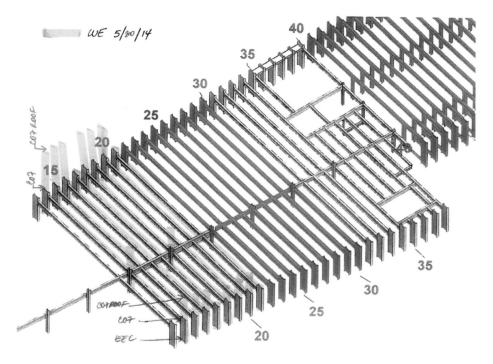

Figure 3.3.23
A marked-up paper print of steel installation at Queens Plaza, tracking up-to-date progress

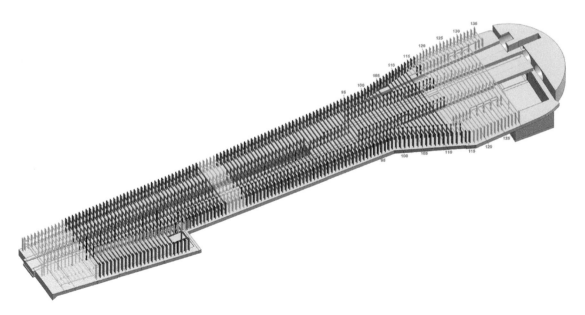

Figure 3.3.24
Updated tracking model of Queens Plaza

image generated from the information model depicting the portion of the facility that was currently under construction. The inspectors used markers to highlight inspected elements. The highlighted images were sent to the VDC team to input into the information model, creating a manual feedback loop between the CM and VDC teams.

The second iteration of the VDC tracking effort was applied to the south contract, which contained large portions of tunnels, cavern elements, and a major ventilation facility. This was the first contract in which tracking was implemented from the award of the contract, enabling the VDC team to participate in the design of the CM tracking workflow, rather than develop a process to support an existing, non-technologically enabled workflow.

A dedicated VDC field coordinator was assigned to the south contract field office to implement and maintain the VDC methodology. The information model for the south contract was designed to integrate the progress tracking data by attaching parameters to the model elements. The VDC team designed the information model with an understanding of the changing nature of construction; they therefore built flexibility into the system, allowing it to evolve and adjust as the tracking effort progressed. The CM team quickly recognized the value of VDC, and the scope of the engagement increased. The information model became a central node of the CM's tracking workflow and the hub for data collection, verification, and visualization of construction operations.

Data visualization based on the information model enabled the team to recognize redundancy of tracked data coming from the field. Due to the manual nature of tracking progress on the construction site, data describing weekly field conditions are easily corrupted, introducing inaccuracy into the reports. Creating a feedback loop between data input and data visualization gave the CM team a quality control mechanism that improved the reliability of weekly tracking reports. Moving forward, the goal of the VDC team was to streamline and automate this process, limiting errors from the manual input of data into the model.

As the volume of tracked elements increased, operating and updating the model became more demanding for the VDC field coordinator. The VDC team created an automated data input solution, which included a custom-designed automation routine to streamline the data input from Excel to Revit. The automated process for interaction between the data and the geometrical elements harvested the power of both Autodesk Revit and Microsoft Excel. Although automation requires a bigger commitment of VDC resources up front, the payout is clear when it comes to day-to-day model maintenance and operation. In the case of the south contract,

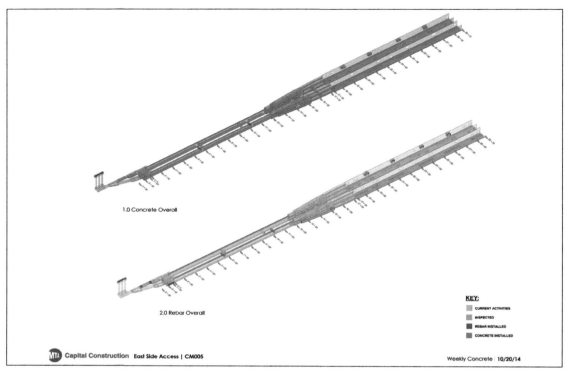

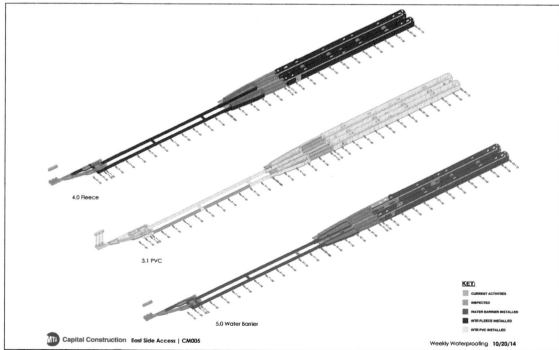

Figures 3.3.25A and 3.3.25B
Typical views of the south contract tracking model

installed components are tracked daily, and data are extracted from the model on a weekly basis to generate overall progress reports for the organization, as well as to assist in the compilation of payment packages to the contractor for completed work.

The third iteration of the tracking effort was implemented on the north contract, which was similar to the south contract in scope. Having already developed tracking methodologies for other areas of ESA, the VDC team had a solid platform and working experience to design the next version of the tracking effort.

One area where the VDC team saw an opportunity for improvement was in the way that information was captured in the field and communicated to the central office. Inspectors responsible for reporting the status of operations to CM personnel relied on paper reports to record their observations. The process of translating the observations to a digital format involved manually inputting recorded data into individual excel sheets, which were compiled by the lead inspector into a master Excel sheet of all inspector reports.

After analyzing existing data-capturing methods, it became apparent that a great level of redundancy could be eliminated by introducing a workflow designed around a web application. The VDC team created a

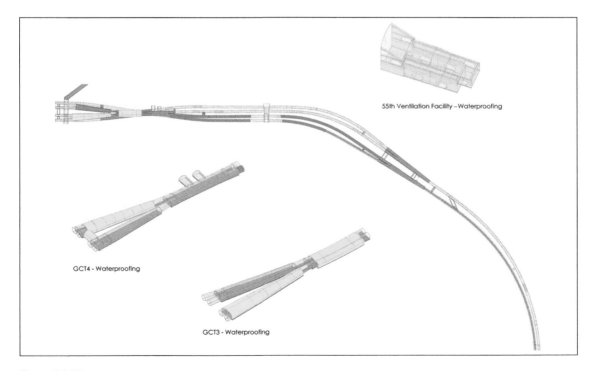

Figure 3.3.26
CM006 tracking model, illustrating waterproofing progress

custom, database-driven web application to augment the process of data collection by inspectors on the job site. At the time of writing, this web-based workflow is going through the testing and implementation phase.

Another addition to the CM workflow was the introduction of a 4-D planning component into the daily operations of the CM's project team. The VDC team provided 4-D training to the VDC field coordinator, extending the VDC capabilities of the CM team. The VDC and CM teams worked in conjunction to develop a 4-D model of the contract based on the Critical Path Method (http://en.wikipedia.org/wiki/Critical_path_method) schedule submitted by the contractor. As this 4-D planning component matured, the standard reporting format evolved to include an image from the 4-D tracking model illustrating the contractually planned work, as well as an image from the updated information tracking model reflecting current progress on the job site.

Contract Model

A contract model is an information model depicting the scope of work of a specific contract. ESA consisted of many contracts, and these information models become essential to understanding the scope of work for each contract. At different stages of the project development, different types of contract models were generated to augment standard workflows. The purpose of the contract models varied from visualization of the contract scope to fully developed VDC contract ecosystems. The most developed contract model included an accurate depiction of existing conditions based on 3-D scan data and LOD 300-level information models depicting the architectural and structural configurations based on 100 percent Design Documents.

The first iteration of the contract model depicted the geometric and spatial configurations of various areas of the ESA project for visualization purposes. The VDC team exported specific contract areas from the master information model as "light models," in Navisworks view-only 3-D format. The "light models" were typically delivered to prospective bidders, accompanied by a package of static images. These assisted the bidders with developing constructability schedules and cost estimates, as well as in presentations to ESA management.

The second iteration of the contract model included providing an information model in Revit format to the systems contractor. The contractor requested the information model, as he saw it as a useful tool to assist his engineering team in configuring the alignment cables. The information model, built by the VDC team, contained dimensionally

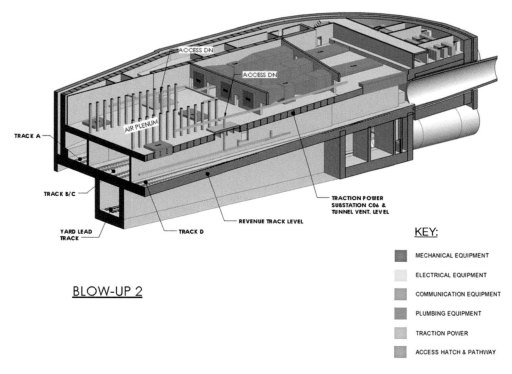

BLOW-UP 2

KEY:

■ MECHANICAL EQUIPMENT

□ ELECTRICAL EQUIPMENT

■ COMMUNICATION EQUIPMENT

■ PLUMBING EQUIPMENT

□ TRACTION POWER

■ ACCESS HATCH & PATHWAY

accurate 3-D structural geometry of the relevant portions of the project. The contractor used this model as a basis to develop a detailed systems model, and he was required to verify all of the dimensional information in the model against design drawings and field conditions.

After the successful use of the contract model in the two cases outlined above, ESA management required the contractor for the concourse contract to develop their own, in-house VDC methodology for the construction of the project. To ensure the success of the effort, ESA's VDC team was asked to prepare several deliverables to provide the contractor with a solid starting point from which to begin their construction modeling.

The concourse level of the project is a connector between Grand Central, various street entrances, and the platforms that house the eight train tracks down in the caverns. It consists of the main circulation corridor and support spaces including ticketing, retail, and other vendor functions. It stretches over eight city blocks and is housed in a former train yard underneath Park Avenue. There are myriad supporting columns throughout the space, and in many areas the clear height to the underside of the slab above is no more than 14 feet. These constraints made the space very limited. Maintaining sufficient ceiling height for the public

Figure 3.3.27
A page from the systems book, illustrating access paths

Figure 3.3.28
3-D view of the concourse, exported from Revit design intent model. The concourse is eight city blocks long

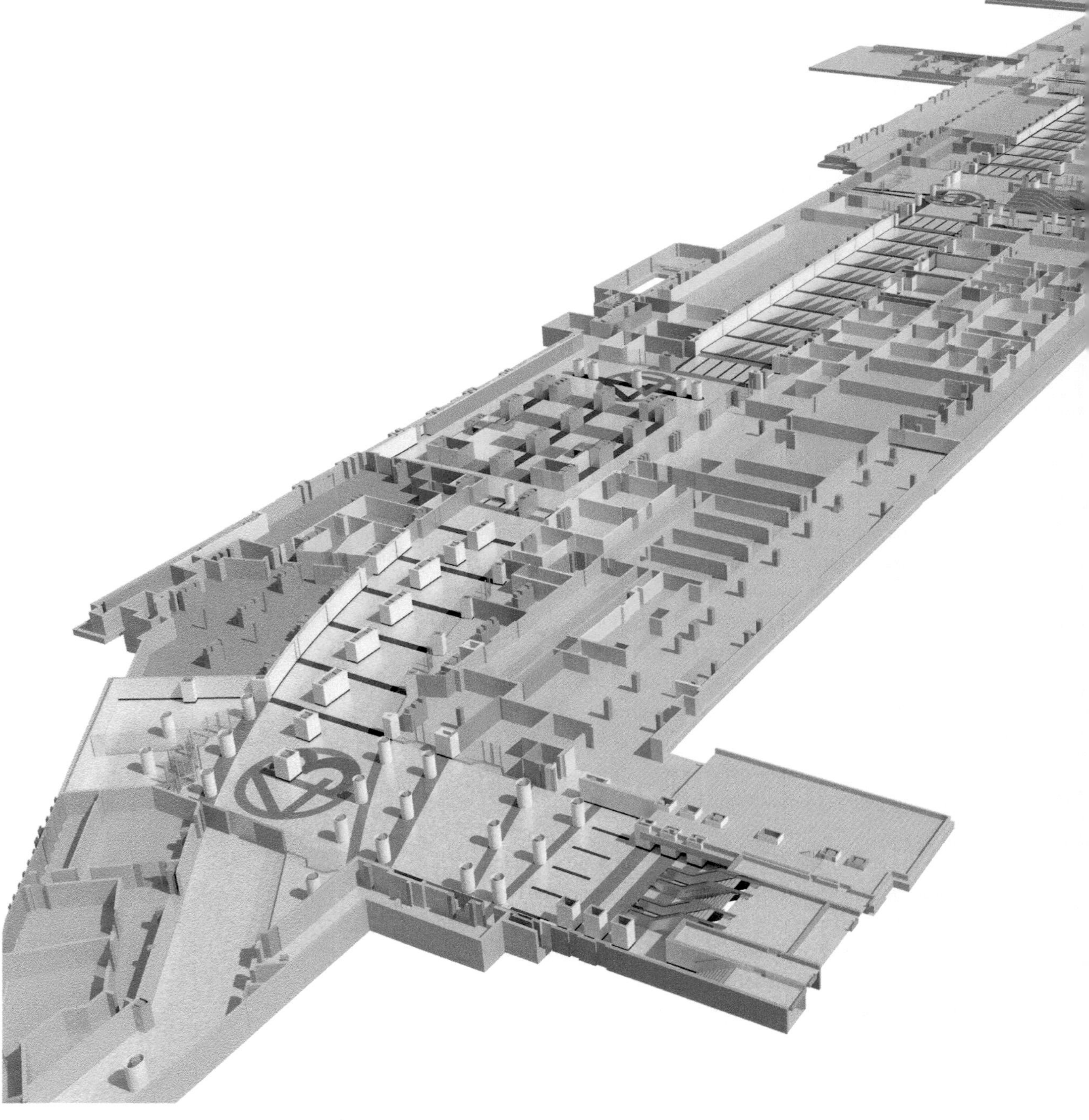

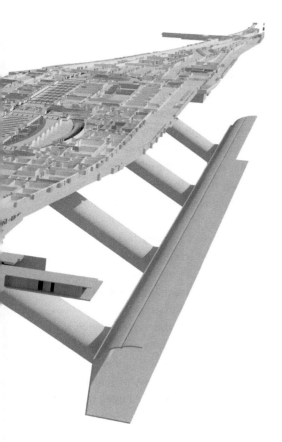

spaces while also fitting all the supporting equipment between the ceiling and the structure above is a challenging task. It was important to understand and plan correct routings before construction started, in order to avoid lowering or redesigning an already constrained space. The proposed design consisted of thousands of hangers from which the finish ceilings would be hung. Due to the irregular nature of the structure above, every hanger location had to be studied in detail.

MTA's CM group, the VDC team, and the MTA design team jointly developed the VDC/ BIM specifications during a series of open forum meetings. As the design team was responsible for the contract documents, it was decided that the design team would author the specifications in collaboration with the other team members. The strategy was for the VDC team to further develop previously created concourse information models for handoff to the contractor.

The package for the contractor included the master model of the entire ESA project, which was accurately positioned in the New York State reference coordinate system. The master model defined global project levels and established a local zero project point, which located the concourse geometry within the reference coordinate system, ensuring the proper placement of all of the models in the greater ESA virtual space. The existing conditions of the entire concourse space, estimated to be 350,000 square feet, were captured using 3-D scanning technology and delivered as an existing conditions information model. The contractor was also provided with an information model of the architectural design intent developed to the LOD 300 standard for the entirety of the concourse from the 100 percent Design Development drawings.

The contractor was then responsible for taking ownership of the supplied models and turning them into LOD 400 standard construction models. This gave them a running start on planning the job and initiating coordination of the different trades. The contract also specified that the contractor was responsible for continuously generating an as-built model, in which COBie facilities management data would be included for handover at the end of the project.

Figure 3.3.29
Members of the Maser team with one
of two scanners, scanning the
concourse

Delivering the contract models allowed the VDC team to embed their VDC standards into the files, including construction tracking parameters and views, which was essential for future collaboration with the contractor. The contractor was responsible for modeling the system runs, so any clearance issues could be identified and rectified prior to installation in the field. The main goal of the effort was to reduce potential change orders that would increase the cost and result in additional delays in the construction schedule.

Maser, the scanning company, had experience with similar large-scale projects and initially proposed authoring the Revit model reflecting the scanned conditions. As the VDC team was so well versed in the project and already had a rough existing conditions model, it was decided that this modeling effort be done by the VDC team using the point clouds as a reference. This turned out to be a far more cost-effective solution. It was also a way of guaranteeing that the model would be generated using the existing modeling standards, so that it would fit into the existing ESA VDC ecosystem.

Since the inception of the VDC effort, the team had been developing the VDC architectural design intent model in stages. The first iteration of the model was simply walls, doors, and structural elements, based on the 2-D CAD drawings provided by the project architects. The goal was to create a 3-D geometry for the purpose of generating 4-D simulation of the build-out. Thereafter parts of the architectural design intent model were further developed to aid in walk-through animations of certain areas. Some of the steel was subsequently modeled as part of an additional planning effort. All the systems equipment and a portion of the pipes and ducts were also modeled previously as part of an installation access study. The rest of the major proposed ducts were added to the model, although this was not part of the original scope of work. A disclaimer was issued on handover, so that the contractor would be clearly notified that not every

Figure 3.3.30A
Image created to show the point cloud model (left side) combined with the generated Revit model of the existing conditions (right side)

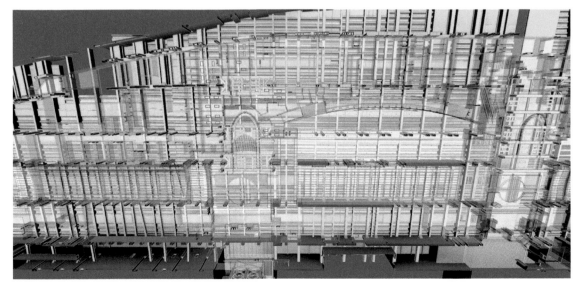

Figure 3.3.30B
Model image showing existing (in orange) vs. proposed ceiling (in grey)

duct and pipe was modeled. This resulted in a set of Revit design intent models that could be populated with as-built facilities data throughout the construction process.

As a final check, the VDC team linked the existing conditions model into the architectural design intent models.

It quickly became apparent that there were many locations where space between the existing ceiling and the proposed design was far more limited than previously assumed. The sections from the 2-D architectural drawing set were depicted in the information model so that the exact discrepancies between the drawings and the existing conditions could be evaluated. The first section showed that there were up to 18 inches less space than was previously assumed. This level of accuracy in the modeling of existing conditions and the proposed design is essential for identifying many problems that otherwise would not be exposed until construction was well under way. It also provided the contractor with an immense advantage in working to deliver the project on time and on budget.

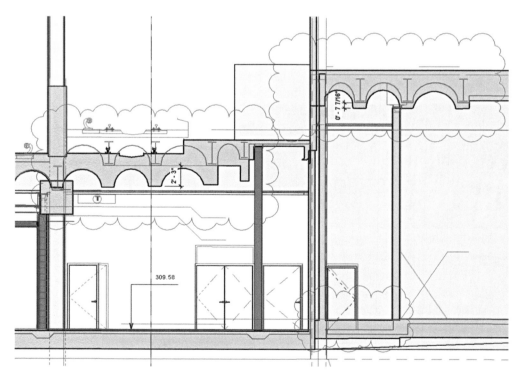

Figure 3.3.31
Section showing ceiling height discrepancy

MODELING FROM POINT CLOUD LASERS

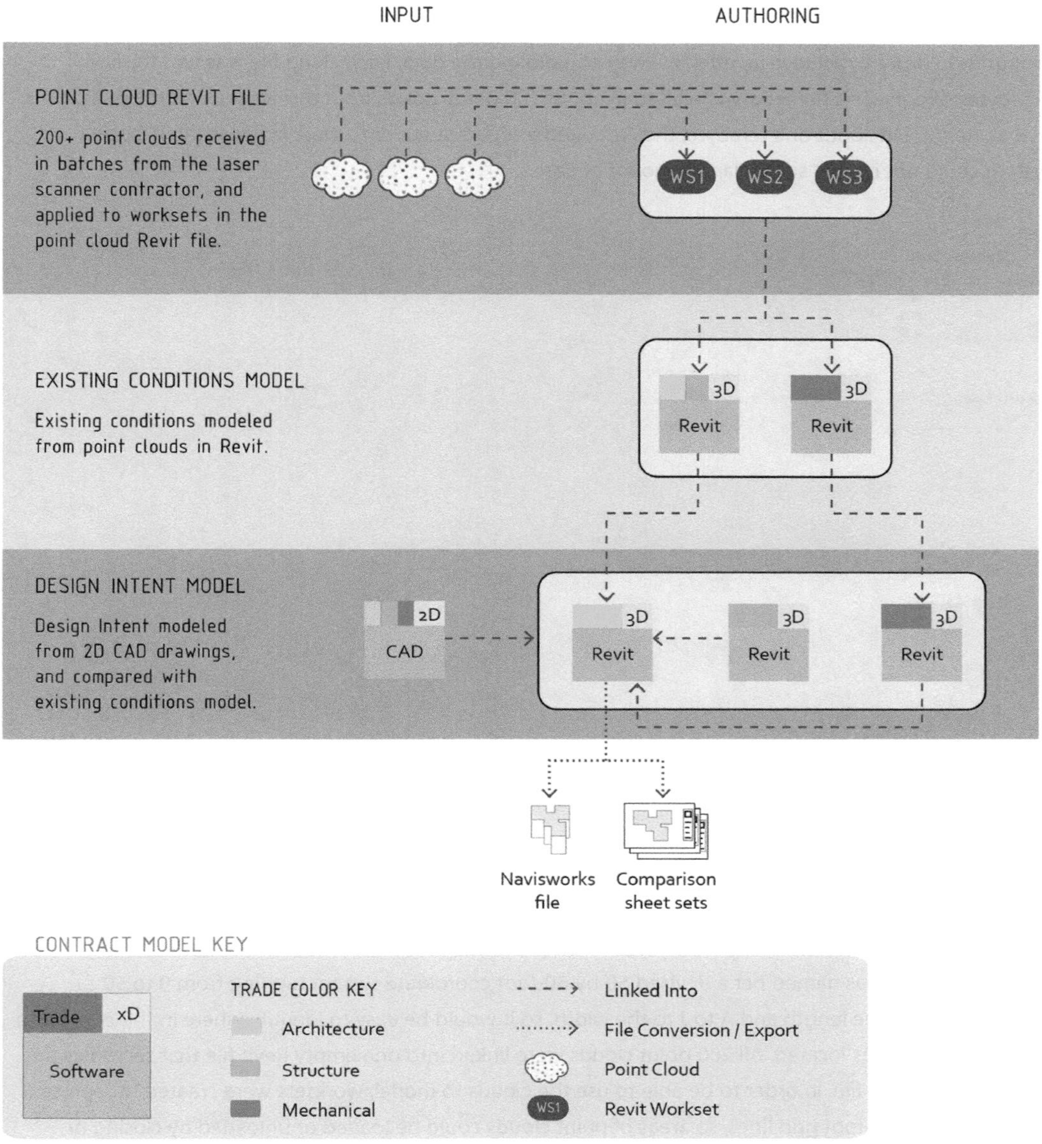

INPUT   AUTHORING

POINT CLOUD REVIT FILE

200+ point clouds received
in batches from the laser
scanner contractor, and
applied to worksets in the
point cloud Revit file.

WS1   WS2   WS3

EXISTING CONDITIONS MODEL

Existing conditions modeled
from point clouds in Revit.

3D Revit   3D Revit

DESIGN INTENT MODEL

Design Intent modeled
from 2D CAD drawings,
and compared with
existing conditions model.

2D CAD

3D Revit   3D Revit   3D Revit

Navisworks   Comparison
file   sheet sets

CONTRACT MODEL KEY

| Trade | xD |
| Software | |

TRADE COLOR KEY

Architecture

Structure

Mechanical

- - - - >   Linked Into

...........>   File Conversion / Export

Point Cloud

WS1   Revit Workset

Figure 3.3.32
Concourse existing conditions scanning: diagram of point cloud model set-up and file organization

Capturing existing conditions using a laser scanner results in point clouds, compilations of millions of points that each have an x, y, z coordinate with a brightness value from the surface. There is no surface geometry in the cloud, only points, so it can only be used as a reference when creating an information model. Point cloud files are very large, so for the concourse project, scan files were broken into 50-by-50-foot grids in order to achieve an effective way of handling the data. Each cloud file was two to six gigabytes of data. This breakdown resulted in about 200 point clouds from the scanner. As the point cloud files alone totaled about one terabyte, they required a separate server to work from, as the IT backup system could not handle such a large amount of data.

Figure 3.3.33
Screen-shot image of the concourse in the TruView web application. The scans were available for the project team's perusal in the form of 360-degree views from each scanning station

Each point cloud was named per a devised 50-by-50-foot coordinate system running from 0 to 50 perpendicular to the length and A to J to the width, so it would be easy to identify where in the concourse level each cloud was located. All 200 point clouds were linked into one empty Revit file that served as the master point cloud file. In order to be able to use the clouds to model, worksets were created to represent ranges of 50-by-50-foot grid lines, so areas of point clouds could be loaded or unloaded by closing or opening worksets locally.

Views in Revit were generated for 1-foot-deep slices in plan and section. These slices provided clear silhouettes of the existing conditions that were used to generate the model. For the plans, cuts were made at almost every foot. The plan and sections were placed on sheets in Revit so that the point cloud could be printed for review if necessary. The point cloud ceiling plan proved to be a useful tool, as it looked like a black and white photograph of the existing concourse ceiling.

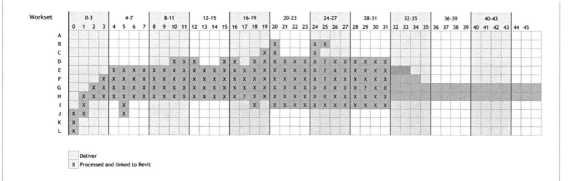

Figure 3.3.34
An image of the Excel file used to track incoming point clouds from the laser scanner contractor. The area of the concourse contract has been divided into a grid of 50-by-50-foot squares. The grey stripes delineate the worksets to which the cloud files will be assigned in the Revit-generated existing conditions file. 'X' indicates that the cloud file for that concourse portion has been received. This image shows that about half the cloud files have been received

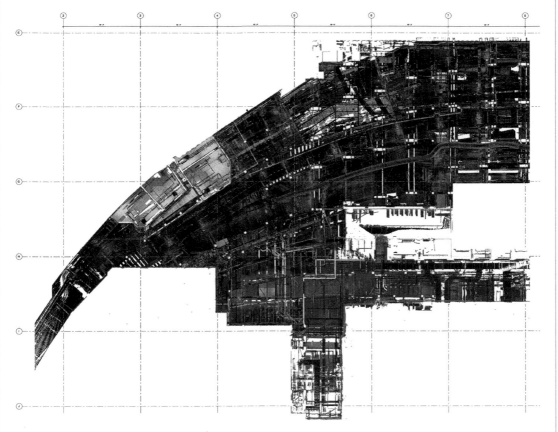

Figure 3.3.35
View of the point cloud ceiling

A set of Navisworks files were also set up so that the point clouds could be navigated in 3-D space. It is possible to do this in Revit as well, but the performance is much faster in Navisworks.

Figure 3.3.36
Point clouds of project viewed in Navisworks. The Navisworks files are distributed to the project team so that they can navigate through the existing conditions

Files previously developed by the VDC department were used for the modeling of the existing structure and architecture. The point cloud Revit files were linked in and used as a reference to adjust the geometry in the files. The big goal was to accurately model both the concrete slabs above and the structural steel that had been recently installed above the open areas, where the large connector tunnels would house escalators leading down to the caverns 150 feet below ground.

The limitation of most modeling software is its inability to capture irregularities of the true conditions. Sagging or bending structural members are a significant concern, as modeling such deviances in geometry is very time-consuming. Much of the original structure contained irregular arches of different sizes, but it is difficult to see geometrical inconsistencies at first glance. The arches were modeled by extruding them from the section views and avoiding deviations of points greater than an inch. Fortunately, the vast majority of existing conditions were within the one-inch tolerance. The contractor would also receive the linked-in point clouds to complement the information models.

The concourse level contains existing conduit runs that had to be modeled accurately in size and location. Some of the conduits were put in place years ago, and some were part of another contract that was just being completed while scanning was taking place. The recently installed systems had already been

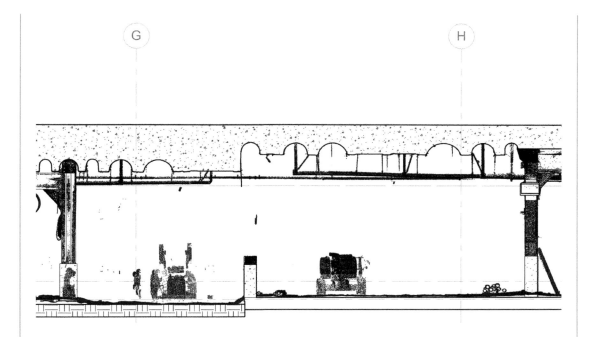

Figure 3.3.37
Section of silhouette of model generated from the point cloud

modeled in the system design intent model, so they were copied into the existing conditions system model and adjusted to the true installation locations.

For the older systems, the team evaluated different automation tools to simplify generating pipes in the model based on the scan data. One such software was a Revit plugin called Imaginit Scan to BIM, which generates pipes by selecting points in the cloud, then calculating the size and location and creating Revit pipes.

EAST SIDE ACCESS: CONCLUSION

This book was completed at the midpoint of ESA's construction. The future development of VDC technologies during the construction of this project are thrilling to imagine. Models will be further linked to the field such that the feedback loop between the model and the site will be almost instantaneous. The VDC effort on ESA has been and continues to be a great VDC laboratory. Initially the project did not include any VDC products, but the use of the models eventually touched many CM-related activities. This was achieved using a small team on a project that employs thousands.

When the first model was created, it was unknown whether the software and hardware were going to be powerful enough to handle the size of the model and the amount of data. To mitigate this fear, the VDC team gradually worked from less to more detail, trusting that Moore's law of exponential technological evolution would keep up with the team's accumulation of data. When the detail level increased, technological advancements worked in our favor, as the latest hardware could process ever-increasing amounts of data.

This project has been plagued with multiple delays and cost overruns since its inception. The implementation of VDC methodologies has, in many ways, revolutionized the understanding of the many thousands of different components and interrelated decisions that make up a project of this size and complexity. The interconnected series of VDC products serves as an efficient information platform, which enables informed decisions to be made before many issues arise in the field—issues that have a tendency to set off delays and cost increases like a line of falling dominoes.

The very fact that projects of this magnitude have been documented two-dimensionally, drawing by drawing, quickly becomes almost unfathomable once VDC tools are introduced and relationships between all parts of the project become clear. Instead of 100,000 drawings, there are 50 interconnected models that serve as a recipe for construction. This is one of the true powers of VDC.

3.4 RFK Toll Plaza Rehabilitation

Robert F. Kennedy (RFK) Bronx Toll Plaza rehabilitation and Manhattan plaza interim repairs is a multi-staged transportation renovation project and one of the final phases of the overall RFK Bridge rehabilitation program. The RFK bridge is located in New York City on Randall's Island, between upper Manhattan and Northern Queens, consisting of two toll plazas and three bridges, which serve as a main artery connecting the boroughs of Manhattan, the Bronx, and Queens. The bridges and toll plazas were originally built in the 1930s, using funds from President Roosevelt's New Deal initiative. The project was championed by controversial "master builder" Robert Moses. It was one of the largest infrastructure projects of its time, even larger than the Hoover Dam that was constructed concurrently. Connected to the Manhattan toll plaza stands the iconic Art Deco building that became Moses' office for the remainder of his career. In 1969, the toll plazas were greatly expanded due to increased traffic, and widening structures were added, nearly tripling the number of toll booths.

Today, about 200,000 vehicles pass through the plaza every day. The roadway is elevated on columns above the island. There are two large toll plazas, one 18-lane and the other 16-lane, one serving Manhattan and the other Bronx–Queens traffic. The scope of the RFK Bronx Toll Plaza project consists of the Bronx toll structures and plaza. The project scope includes total replacement of the roadway, including replacement of steel; repair and some replacement of the concrete piers; and replacement of the existing toll booths. The Manhattan plaza portion of the project is much smaller in scope, consisting of repairs of the supporting structure underneath the toll plaza. The project is scheduled to last five years and contains multiple phases, so an acceptable flow of traffic could be maintained throughout the duration of the project. The overall Manhattan plaza is to be replaced as a separate project at a later date. The project to renovate the toll plaza and roadway was programmatically and structurally complex, as it applies twenty-first century engineering solutions to structures built in the 1930s and 1960s.

Figure 3.4.1
RFK Toll Plaza site rendering

Bronx Plaza Construction Cost:	$210 million
Location:	Randalls Island, NY
Duration:	2014–19
Type:	Transportation
Client:	Triborough Bridge and Tunnel Authority
VDC Fee Structure:	Included in construction management services
Contract Type:	Design, bid, build
VDC Focus:	BIM civil and structural modeling, review, model integration, and tracking and reporting

VDC Technologies

Information Model Authoring:	Autodesk Revit
Coordination, Clash Detection:	Autodesk Navisworks
Tracking/Planning:	Revit, Excel, Navisworks
Project Information Management:	Newforma
Visualization/Animation:	Lumion, Adobe Premiere

Figure 3.4.2
Site image from Google Earth

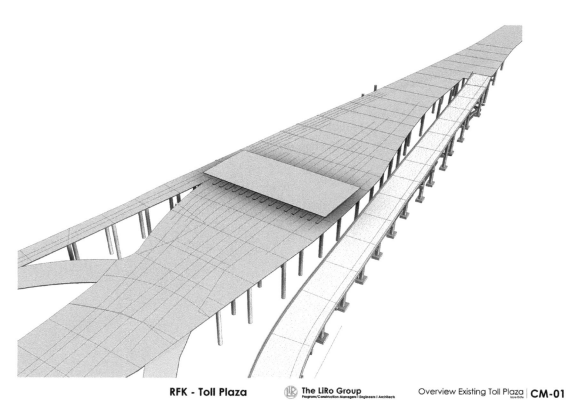

RFK - Toll Plaza   The LiRo Group   Overview Existing Toll Plaza | **CM-01**

Figures 3.4.3A and 3.4.3B
Overview of the Bronx Toll Plaza, existing design and new design

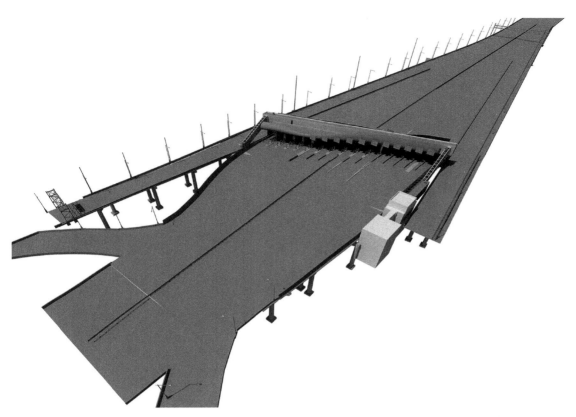

Figures 3.4.3C Rendering of completed Toll Plaza

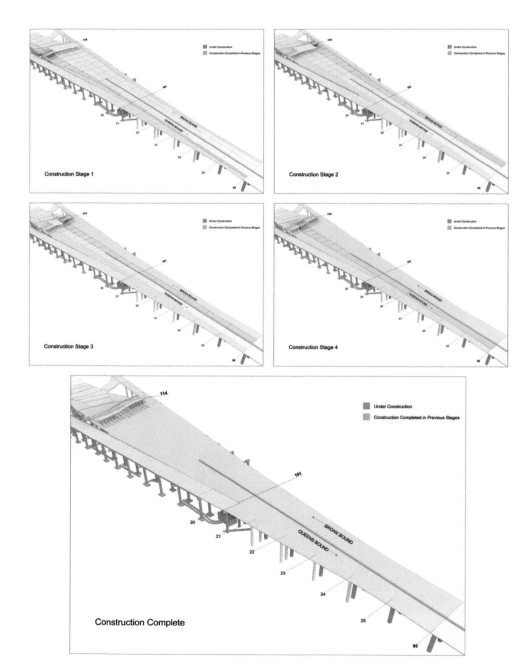

Figure 3.4.4
Project phases, as presented in LiRo's initial proposal

PROJECT BACKGROUND

The most challenging part of the project was maintaining traffic flow while performing complex construction procedures. Careful phasing of the construction was necessary to prevent undue disruption of traffic.

When LiRo prepared its presentation for the opportunity to manage the construction of the project, there were no drawings provided to the VDC team. Drawings of the toll plaza could only be accessed and reviewed in MTA's main library in Manhattan for a few hours at a time. By tracing the critical drawings on tracing paper, the VDC team

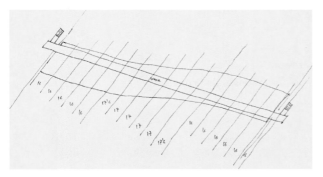
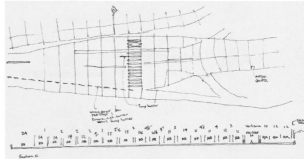

Figure 3.4.5A and 3.4.5B
Scans of the traced toll plaza drawings

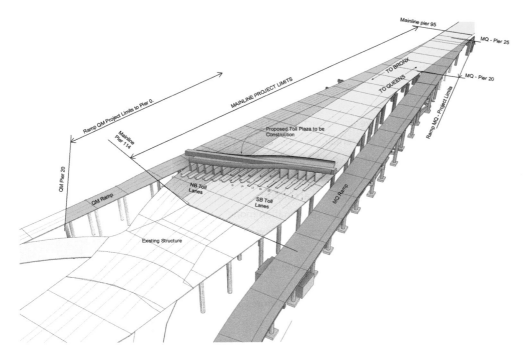

Figure 3.4.6
Rough staging model

generated a schematic model, which was used to illustrate project understanding and how the phasing work could be interpreted. This rudimentary model was used in the proposal, as well as during the live proposal presentation, as a tool for the CM team. The project was subsequently awarded to LiRo.

GOALS

- Use the information model as a tool to understand the existing site conditions.
- Use the information model to plan the sequence of the work.
- Use an information model-driven database for tracking progress and inspections.
- Document the routing of the surrounding existing services.
- Clash detection and systems coordination.
- Use the information model for constructability review.
- Generate an as-built information model for future operations and maintenance.

SUCCESSES

- Generated a conceptual construction information presentation model based on sketches traced from paper drawings.
- Used Revit to accurately model infrastructure, like roadways.
- Performed constructability review of the completed project.
- Fused the toll booth design intent model provided by the design team with the plaza model created by the LiRo VDC team.
- Performed systems coordination between the existing and proposed systems.
- Provided a solid foundation to subcontractor for coordination modeling and the coordination process.
- Developed tracking methodologies tailored to civil and structural repairs.

CHALLENGES

- The design team modeled the above deck structure including the toll booths and adjacent toll utility building. The VDC team modeled the remaining elevated plaza. The modeling of the supporting structure and roadway was labor intensive, as it had to be converted from a large amount of old hand drawings and CAD documents.
- The provided design model was not organized according to a clear BIM standard. It was quite an effort to organize them before linking them to the coordination model.
- Set-up of the tracking process required a thorough review of how tracking is done on an infrastructure project.
- Understanding the guiding principles of construction when integrating new roadway systems with previous construction from the 1930s and 1960s.
- Creation of an information model of the complex curvatures and slopes of the roadway, using a software geared toward buildings.

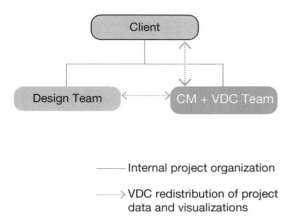

Figure 3.4.7
Organizational diagram

─────── Internal project organization

············> VDC redistribution of project
data and visualizations

VDC STRATEGY AND DELIVERY

Project Set-up

Initial work involved repairs to the support structure of the Manhattan-bound toll plaza. In this area of the project, the model was used to plan, track, and visualize the repair statuses on one flat level. Consequently, the model was created as a simplified flat 2-D plane, as the 3-D geometry of this area was not essential for planning and tracking. It was more efficient to model in 2-D, but still use the BIM data to drive and visualize the progress.

The second phase of work was in the Bronx Plaza, which included the complete replacement of the toll plaza. The toll booth structure had already been modeled by the architecture and engineering team. Their design intent information model included the architecture, MEP, and structure of the project categorized by existing, demolition, and new phases. The roadway segments beyond the toll plaza and underneath the toll booths were not part of the designer's model. The VDC team's first task was creating a coordination model of the complete toll plaza, in order to analyze the entirety of the toll complex, including all the existing components and services that would impact the planning and execution of the project.

Stage 1: Manhattan Plaza Tracking Repairs

The Manhattan-bound toll plaza portion of the project consisted of repairing the supporting structure underneath the toll plaza. Different structural members required various types of repairing operations. The design drawings described approximately ten types of repairs, each with a unique identifier. The VDC team designed a system to track and visualize

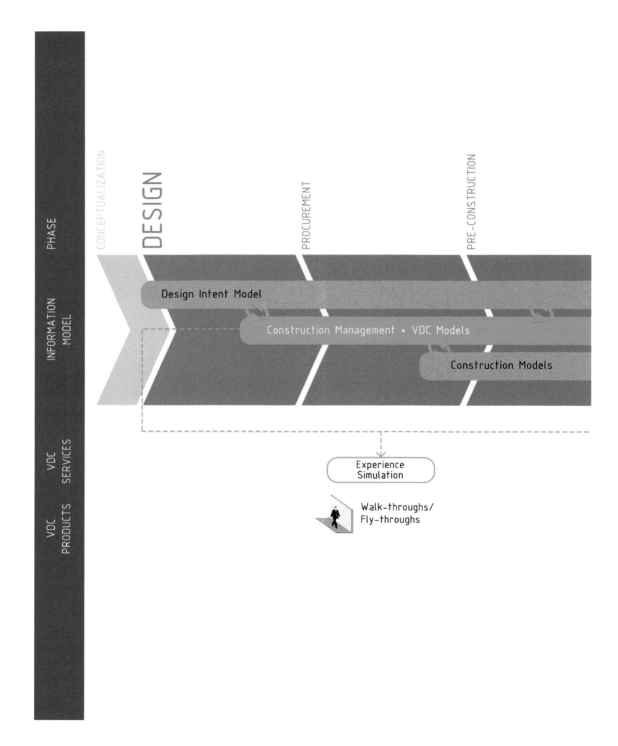

Figure 3.4.8
RK-65 products and services timeline

RK-65 SERVICES + PRODUCTS DIAGRAM

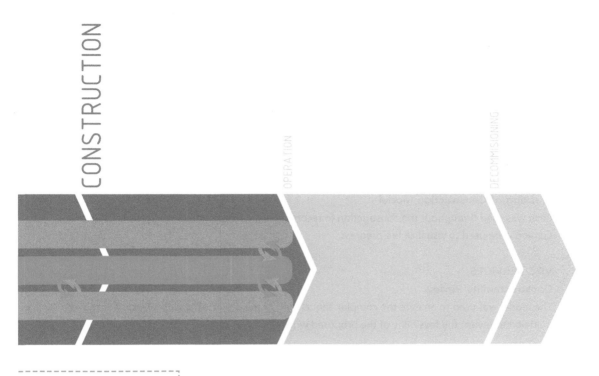

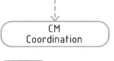
CM
Coordination

 Phasing
drawings

 Distribution
models

 Virtual
Requests for
Information

 Tracking
sheets

 Tracking
report

VDC PRODUCTS

Schematic Construction Presentation Model

A model that was not dimensionally accurate, used to communicate project scope and the phasing of construction. Images from the model were included in the CM bid documents and also used during the live presentation.

Coordination Model

A central information model, broken down into the six phases of construction. It held the existing, demolition, and new components of the structural, architectural, and systems sub-models. Managed by the LiRo VDC team, and used for constructability review and coordination.

Distribution Model

Current Navisworks version of the coordination model.

Tracking of Construction Model

Revit was used throughout the construction to record weekly construction process. Navisworks and Lumion were used to visualize the progress.

VDC SERVICES

Constructability Review

The model was used to analyze the complex sequencing of the project. The existing conditions were modeled to review the feasibility of the proposed work.

Tracking

The model was connected to a tablet-based interface which was used to collect data from the field, updating the model to generate and visualize the progress of installation of interconnected building systems.

Clash Detection

Clash detection was run between the systems in the new toll plaza as well as underneath the roadway.

repair procedures, and created a 2-D information model that showed each tracking type. The benefit of creating a 2-D information model over simply using the 2-D drawings was that this flat model retained all of the database functionality of a 3-D information model, just without the spatial geometry. Data were assigned to areas of 2-D geometry, and parameters were used to produce color-coded drawings to visualize the repair statuses.

Each repair location consisted of up to six repair steps that needed inspection upon completion. Each step was assigned an item number which contained information about the repair, such as description,

RK-65 CONSTRUCTION MANAGEMENT COORDINATION STRUCTURE

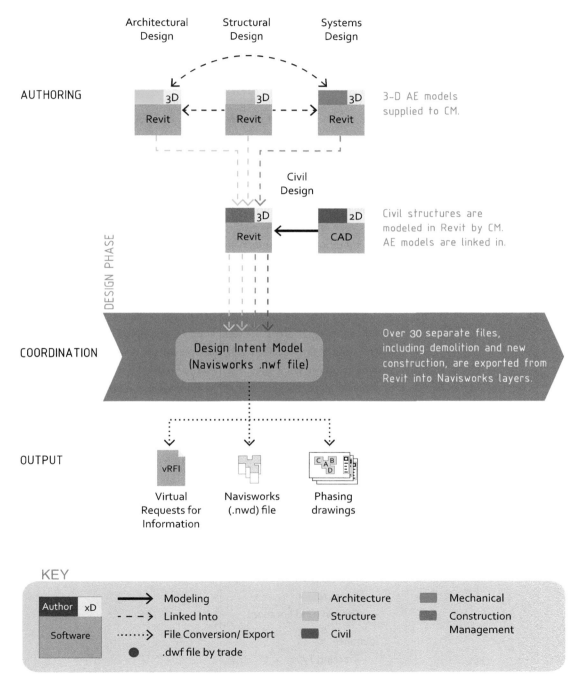

Figure 3.4.9
Construction management coordination structure diagram

quantity, status, data, and inspector name. There were over 600 repair locations, resulting in about 3,500 repair steps, all of which were individually depicted and tracked in the information model. The first step of operations was inspecting all of the elements contained in the scope of the contract. Critical and high-priority repairs were flagged as either red or yellow. Red-flag operations needed immediate attention, because of the severity of the structural integrity of the elements. Yellow flags were lower priority, but still needed to be addressed prior to the remaining repairs.

Using a custom-built automation routine, Excel data tables were generated from the information model and populated with the statuses of all of the structural elements by inspectors using tablet computers in the field. In the Excel sheet, there were tabs set up for each type of repair to facilitate quick access for the inspector. Each repair location was represented in the information model using symbols from the CAD drawings, and made into dynamic parametric objects that contained localized data about the repairs.

A location coding system was driven by the type and location of each repair. Four types of data were used to determine location: span number, pier location, girder number, and grid line. As each type of repair was associated with a geometrical symbol, individual repairs could be read and color-coded on the tracking plan based on their status. To signify the number of repair steps, the team adopted a graphical language, which consisted of dots located below each parametric repair symbol. Once a repair step was completed, a dot was converted from color red to green. When all steps were completed, the main repair symbol was shown in green.

When data were synced on field tablets, the progress of operations on the tracking plan was visualized in the information model. PDFs of the plans indicating progress were printed from the information model in various formats.

The following data were required for maintaining accurate tracking in the information model:

Location
- Girder Location
- Pier Location
- Room Name
- Section (Grid Location)

Identifier
- Tracking Type
- Order

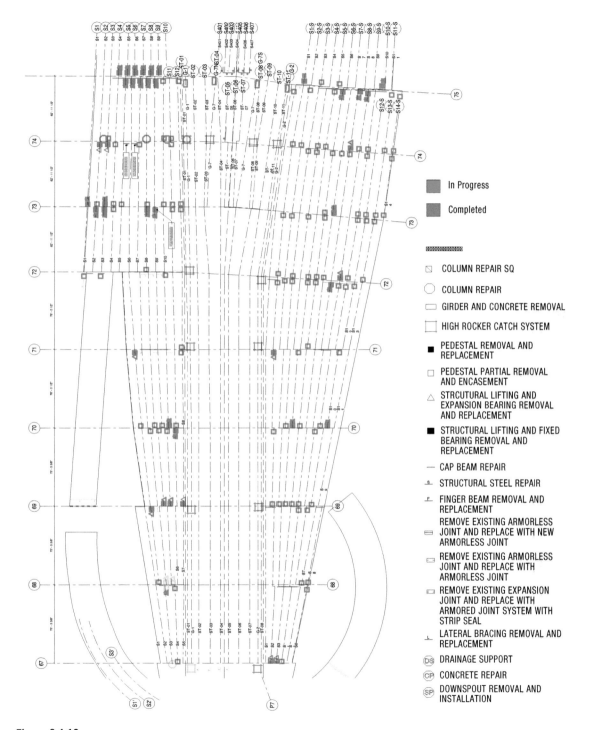

Figure 3.4.10
Manhattan Plaza key plan

RK-75/GFM 509, W.O.1
Construction Management Services for the
Miscellaneous Structural Repairs of Manhattan Plaza at the RFK Bridge

⚠ STRUCTURAL LIFTING AND EXPANSION BEARING REMOVAL AND REPLACEMENT   Item
(DWG. NOS. S-019 AND S-020)   *If Applicable

Date	Status	IDR	*Flag #	Item No.	Description of Work	N/A	Location				QTY.	Unit	Remarks
0						0	NW	P69	SPAN 69N	S3			
0				585.01	Structural Lifting Operations - Type A	0	NW	P69	SPAN 69N	S3		EA	
0				589.5200	Removal of Existing Steel - Bearings	0	NW	P69	SPAN 69N	S3		EA	
0				565.1321	Type S.S. Expansion Bearing (All Load Ran	0	NW	P69	SPAN 69N	S3		EA	
						1	NW	P69	SPAN 69N	S3		EA	
						0	NW	P69	SPAN 70N	S4			
0				585.01	Structural Lifting Operations - Type A	0	NW	P69	SPAN 70N	S4		EA	
0				589.5200	Removal of Existing Steel - Bearings	0	NW	P69	SPAN 70N	S4		EA	
0				565.1321	Type S.S. Expansion Bearing (All Load Ran	0	NW	P69	SPAN 70N	S4		EA	
						1	NW	P69	SPAN 70N	S4		EA	
0						0	NW	P69	SPAN 70N	S5			
0				585.01	Structural Lifting Operations - Type A	0	NW	P69	SPAN 70N	S5		EA	
0				589.5200	Removal of Existing Steel - Bearings	0	NW	P69	SPAN 70N	S5		EA	
0				565.1321	Type S.S. Expansion Bearing (All Load Ran	0	NW	P69	SPAN 70N	S5		EA	
						1	NW	P69	SPAN 70N	S5		EA	
0						0	NW	P69	SPAN 70N	S7			
0				585.01	Structural Lifting Operations - Type A	0	NW	P69	SPAN 70N	S7		EA	
0				589.5200	Removal of Existing Steel - Bearings	0	NW	P69	SPAN 70N	S7		EA	
0				565.1321	Type S.S. Expansion Bearing (All Load Ran	0	NW	P69	SPAN 70N	S7		EA	
						1	NW	P69	SPAN 70N	S7		EA	
0						0	NW	P71	SPAN 71N	S2			
0				585.01	Structural Lifting Operations - Type A	0	NW	P71	SPAN 71N	S2		EA	
0				589.5200	Removal of Existing Steel - Bearings	0	NW	P71	SPAN 71N	S2		EA	
0				565.1321	Type S.S. Expansion Bearing (All Load Ran	0	NW	P71	SPAN 71N	S2		EA	
						1	NW	P71	SPAN 71N	S2		EA	
0						0	NW	P74	SPAN 74N	S2			
0				585.01	Structural Lifting Operations - Ty	0	NW	P74	SPAN 74N	S2		EA	
0				589.5200	Removal of Existing Steel - Bea	0	NW	P74	SPAN 74N	S2		EA	
0				565.1321	Type S.S. Expansion Bearing (	0	NW	P74	SPAN 74N	S2		EA	

Figure 3.4.11
Excel sheet view

Descriptors
- Flag Number
- Item Number
- Description of Work
- Quantity
- Unit (for Quantity)
- Comments

Stage 2: Bronx Plaza: Constructability and Planning

A comprehensive information model was created to manage the complexity of the Bronx Plaza portion of the project, as this stage included multi-phased demolition, construction of new roadways, and toll booth renovations. The information model needed to contain not only the geometry of the new construction, but also the existing conditions relating to the project. Modeling the existing conditions for the entire plaza was essential to generate a comprehensive view of all the systems that were connected to existing facilities underneath the plaza. The area of work included roughly 1,500 feet of roadway. The new toll plaza was to be positioned differently than the old, on a diagonal instead of running perpendicular to the roadway. This design required the construction of a new support structure underneath the roadway and new penetrations to the existing supporting infrastructure. Some of the existing support

Figure 3.4.12
Manhattan Plaza repair symbol in
Revit. Parametric objects, that appear
as graphic symbols, were placed on
plan on specific locations that required
repairs. Colored dots located below
each symbol (here, a triangle)
represented the status of a repair task.
When the symbol is updated to green,
the repairs on its location are
complete

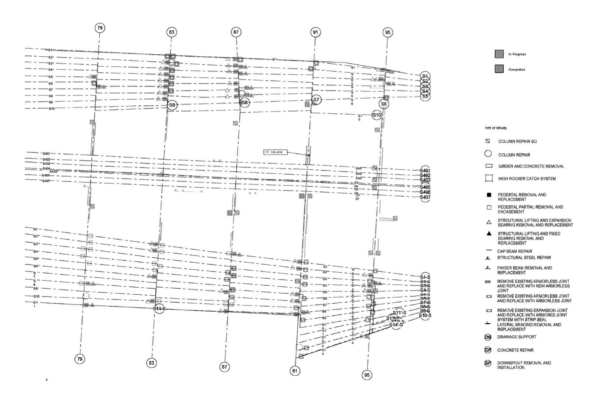

Figure 3.4.13
Tracking repair tasks are indicated with icons that are color-coded with either red or green, to represent if tasks are
incomplete or complete respectively

BREAKDOWN OF RK-65 MODELS: Existing and New Design Intent

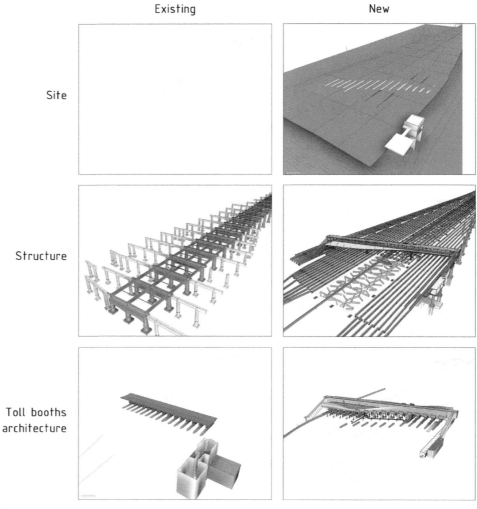

Figure 3.4.14
Breakdown of existing and new models

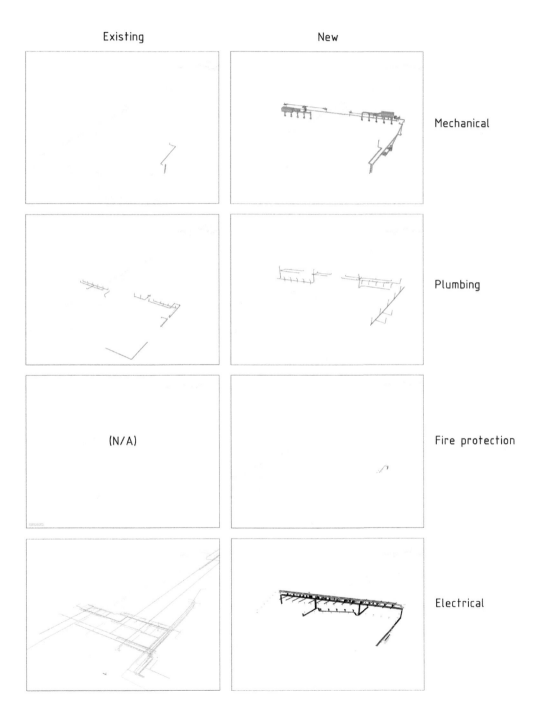

Existing

New

Mechanical

Plumbing

(N/A)

Fire protection

Electrical

Figure 3.4.15
Coordination model exported from Revit

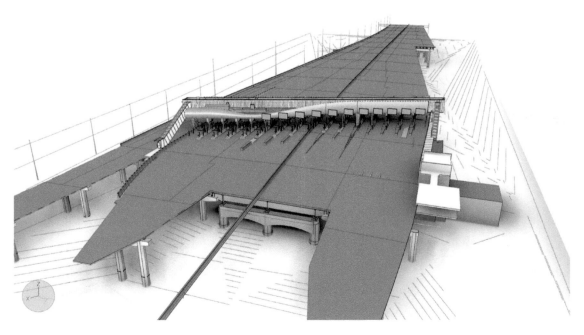

Figure 3.4.16
Coordination model, showing systems, exported from Navisworks

systems had to be kept operational while work was taking place on the toll booths and roadway; therefore a complicated phasing plan needed to be designed and applied to the schedule. The field engineer documented all the existing infrastructure, which was thereafter incorporated into the coordination model.

The VDC team received PDF drawings of the plaza at the commencement of the project, and updated the presentation model to an acceptable level of accuracy in order to create a usable coordination model for the project. The design team used Revit to model the area of the toll booths, showing both existing and new architectural and structural systems. The MEP portion of the information model primarily contained the new design and very little of the existing systems. The VDC team cleaned up and organized these files according to the LiRo VDC standards, and merged them into the coordination model. After verification, the existing conditions were modeled and linked to the coordination model. All the systems were designated with sub-systems color-coded based on the VDC standards. The resulting coordination model contained the

Figure 3.4.17
View of a clash in Navisworks and corresponding plan view with an arrow indicating location of clash in model

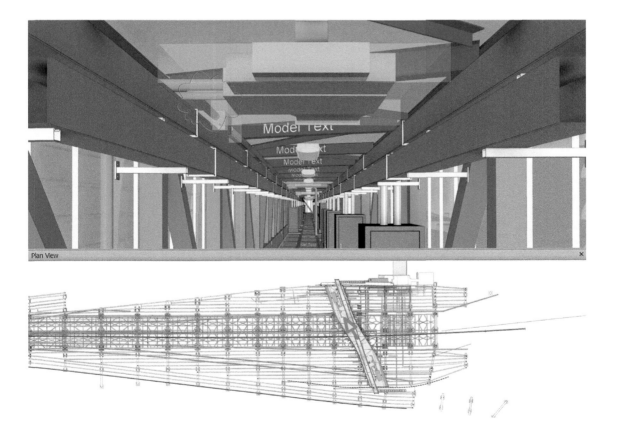

Plan View

Figure 3.4.18
Location tags on columns in the 3-D
model

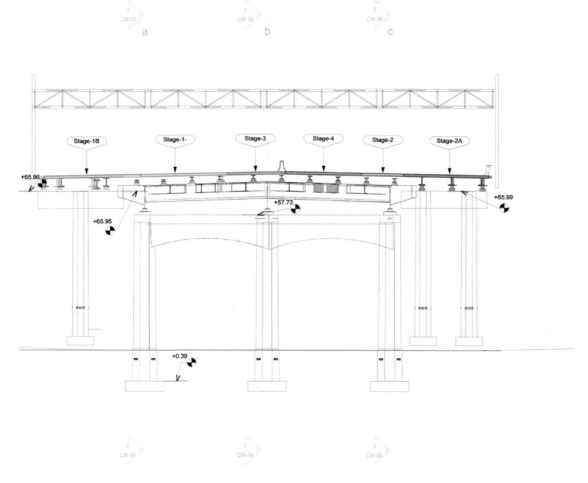

Figure 3.4.19
Coordination sections showing the different construction stages

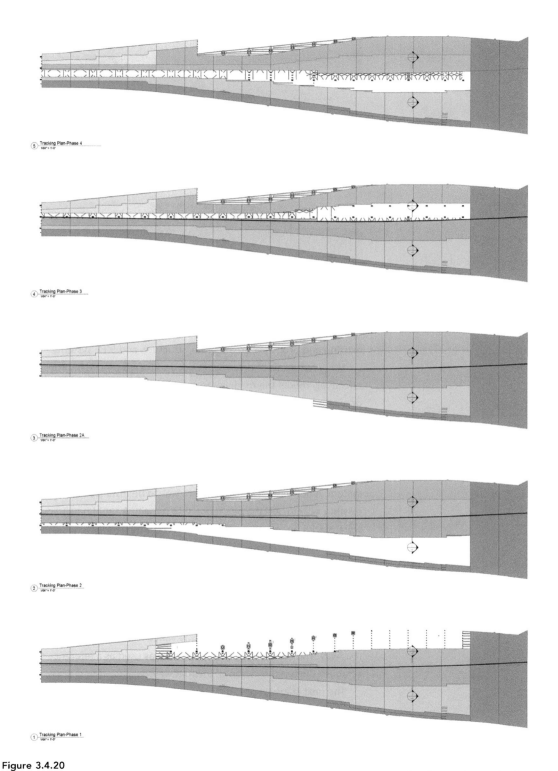

5 Tracking Plan-Phase 4
1/64" = 1'-0"

4 Tracking Plan-Phase 3
1/64" = 1'-0"

3 Tracking Plan-Phase 2A
1/64" = 1'-0"

2 Tracking Plan-Phase 2
1/64" = 1'-0"

1 Tracking Plan-Phase 1
1/64" = 1'-0"

Figure 3.4.20
Coordination plans showing different construction stages. The stages are color-coded so that they can be easily identified across plans, sections, and perspective images

Figure 3.4.21
Filter view

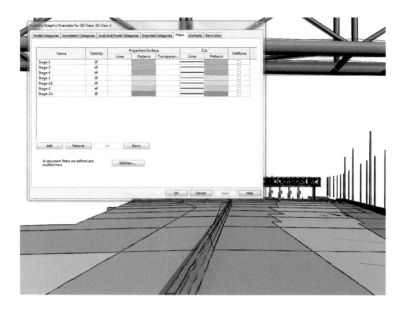

geometry of the existing site conditions, the proposed site design, and the existing and proposed architecture, structure, and MEP systems.

Many hundreds of drawings needed to be referenced in order to effectively model the roadway and the supporting structure. As many drawings were not available in CAD, a considerable amount of time was spent modeling based on dimensional information from drawings and structural schedules. The construction of the elevated road for the original main line consisted of the rehabilitation of existing concrete piers, with new bearings supporting the existing massive steel beam and girder system. A new lateral bracing system was to be installed to replace the existing system and the beams were to remain. A set of new pedestals were placed on top of the beams to support the stringers, which in turn supported the roadway. The height elevations of each pier were modeled based on information from the drawing sections.

For the east and west widening structures, which were constructed in the 1960s, the steel was to be completely replaced. This made them quite easy to model. The hardest part was modeling the roadway, as it sloped away from the center of the driveway varied in elevation along the roadway. The elevation of each pedestal was used as a reference to establish the exact location of the roadway.

The next stage was labeling important components with location tags. The CM team wanted to be able to see the location of columns and steel in the 3-D model. As tags cannot be added to the 3-D Model in Revit, 3-D model text was added to each family in Revit. This model text was then driven by the location parameters set-up for tracking. This is an effective way of making a tag that shows up in 3-D.

Drawing sheets were created in the coordination model which described in plan and section the relationships between the existing, demolition, and construction elements.

Once most of the relevant geometry was modeled, it was time to break the project up into phases. The phasing tool in Revit is quite limited when dealing with multiple phases. In a Revit view, it is possible to show the previous phase or next phase, but there are limited controls for visualizing multiple phases differently in the same view. The VDC team developed a workflow to manage visualizing multiple phases by creating a custom phase parameter in which view filters could be used to control the visibility of any phase as needed. For example, all phases could be displayed in different colors or only the completed phases could be shown green.

The existing conditions and the phases of new construction for each trade were exported as individual models from the coordination model, and assembled in Navisworks for visualization and clash detection. There were many interferences found between the different trades design models, as well as between the existing conditions and proposed work. All the clash locations were saved as views and named according to a naming convention that indicated the location and type of issue. Some issues were discovered by simply "walking" through the model. The route of each pipe was followed to ensure that it was complete and connected correctly.

CONCLUSION

Road and bridge renovation work are construction typologies that require detailed planning, as they deal with conditions that are under constant usage and cannot be interrupted. It requires thorough understanding of the relationships between the phases of work, and the different systems that support it. On the RK-65A Bronx Plaza project, construction documents did not show many of the existing conditions into which much of the new systems had to be tied. The VDC team added these conditions, to create a comprehensive coordination model. Using the coordination model as a tool to plan and communicate the strategy of construction proved extremely useful for this project. At the time of writing, good BIM civil authoring software has yet to be developed. As the LiRo VDC team has extensive knowledge with using Revit, they were able to push the program to its limit in order to effectively describe the complexity of roadway surface.

3.5 Adult Behavioral Health Center at the Bronx Psychiatric Center

Figures 3.5.1A–C
Photographs of the BMHR adult building under construction

The Adult Behavioral Health Center is part of a major capital modernization project at the Bronx Psychiatric Center (BPC) which includes the construction of the Children's and Central Services Buildings, Transitional Living, and Crisis and Studio Apartment Residences (Residential Village).

The 180,000-square-foot Adult Behavioral Health Center is the first new adult inpatient facility designed and built for the New York State Office of Mental Health (OMH) by the Dormitory Authority of the State of New York (DASNY) in nearly two decades. The hospital campus is the DASNY's first project to utilize BIM from the design through construction, and also one of the first projects in the region to do so.

The building is a four-storey structure in a radial layout with patient rooms organized in four wings. It is the main facility on the Bronx Mental Health Hospital campus.

Construction Budget:	$77 million
Location:	Bronx, NY
Duration:	2010–2015
Type:	Hospital
Client:	DASNY/Dormitory Authority of the State of New York/New York State Office of Mental Health (OMH)
VDC Fee Structure:	Included in Construction Management services
Contract Type:	Design, Bid, Build
VDC Focus:	VDC Construction Management, tracking and reporting
VDC Technologies	
Information Model Authoring:	Autodesk Revit
Coordination, Clash Detection:	Autodesk Navisworks
Tracking/Planning:	Revit, Navisworks
Visualization/Animation:	Lumion, Adobe Premiere

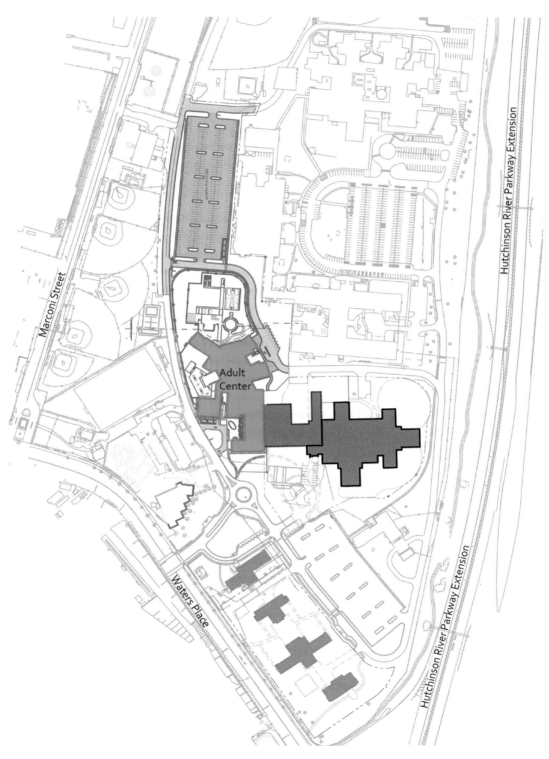

Figure 3.5.2
Map of project

SUCCESSES

- Reviewed and developed quality control mechanisms of the information models.
- Managed the handover of the coordination model, so that the contractor could get it up and running in a timely fashion.
- Developed a tracking method to monitor and communicate construction progress on a week-by-week basis. Incorporated automation tools into Revit to streamline the process, so that information could be captured on Friday and used on Monday mornings for progress meetings.
- Worked with the AE team and client to develop VDC specifications.

CHALLENGES

- Steep learning curve for the construction team, as the Adult Building was a pilot project for VDC and most of the stakeholders were VDC novices in terms of project documentation, coordination, and construction. VDC team expended major effort early on to prove value.
- Inconsistencies between the project coordinate systems used for modeling by the contractor and by the VDC team created unnecessary challenges in comparing the design intent and construction models.
- VDC implementation guidelines were not yet fully developed for the Facilities Management (FM) model during the planning stages resulting in a systems model that was not built for transferring into a facilities management software. This prevented the direct import of the construction model into the facility management software, resulting in needless manual and time-consuming data transfer.
- Enforcing all of the VDC participants to use agreed project standards. Managing the contractor and subcontractor in order to maintain consistent workflows as far as naming conventions and clash recording was time consuming. Failure to adhere to standards made the statistical recording of progress more difficult.

PROJECT BACKGROUND

The LiRo VDC department was formed in 2010, and the Bronx Hospital Adult Building was one of its first projects. Because the mandate to implement VDC came from the client, DASNY, the project team made a concerted effort to engage with VDC processes and technologies from design to construction, despite being new to the concept of VDC. As it was a pilot project for DASNY in using VDC in design and construction, everyone on the project team learned a great deal in terms of how VDC should be applied as a Construction Management tool. DASNY was well intentioned in initiating the VDC effort; a major lesson learned on this project was the importance of involving the experienced VDC team from the beginning in order to define the deliverables correctly and to craft the VDC process to achieve the defined goals. Vague or overstated

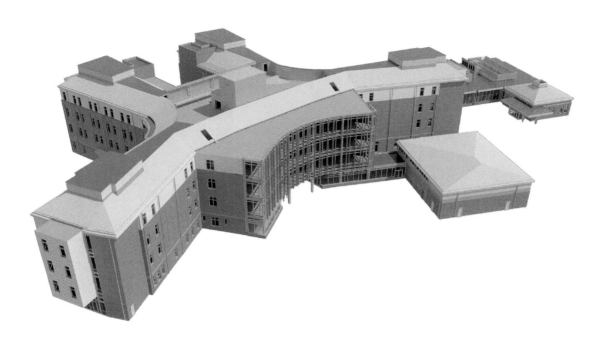

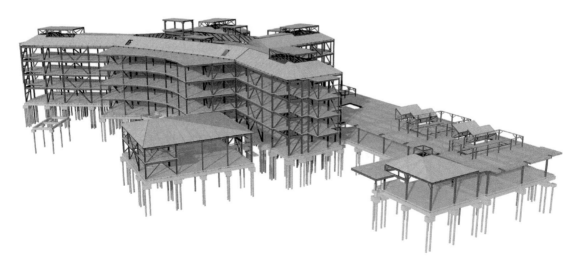

Figure 3.5.3A–B
3-D images of Revit model

Figure 3.5.4
Diagram showing project team
organization

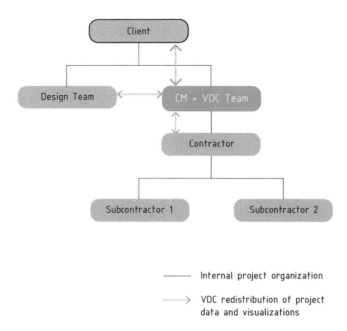

requirements in the contract can cause confusion and lost opportunities as
the project progresses.

The LiRo Group was selected as the construction manager for the
Adult Behavioral Health Center project. LiRo's VDC department was
responsible for working with the construction management team to
observe the implementation of VDC during the design and construction
phases of the project, pilot new VDC processes, and educate and make
recommendations to the team for increasing engagement with VDC.
The LiRo VDC team was engaged during the pre-construction phase and
assisted with the project through construction.

The project's design intent information model was created by the
architect and coordinated by the contractor. The VDC team's role was to
closely observe and manage the VDC process during the pre-construction
and construction phase. This included enforcing VDC standards and
protocols. The model was used extensively for tracking weekly
construction progress.

An additional deliverable specified in the contract documents was an
as-built information model containing building data, which was to be
transferred into the Facility Management software in order to produce
a facilities management model. However, this was not delivered, as the
specifications defined by the project team at the beginning of the project
were not specific enough about what data needed to be attached to the
as-built information model as it was generated during construction.

A detailed description of the FM data needs to be specified at the commencement of the VDC process so that both the design and contractor teams can gradually populate the data throughout the process.

Design Phase

Starting in 2010, the architecture and engineering team developed a design intent information model to LOD 300. This model was used to generate construction drawings that were the legal basis for the bidding process. The design intent model transitioned from the design phase to construction, and the architecture and engineering teams continued to update the model to include changes and sketches generated from construction coordination. One limitation of the model for construction coordination was that when the model was started in 2010, Revit did not have the capability to show insulation of ducts or pipes. This capacity was added in later versions of Revit. Not having the insulation modeled sometimes made it challenging to coordinate the ducts and pipes for clearances issues.

During the design phase, the design team managed the coordination process and ran clash detection in order to provide a substantially clash-free model. The VDC team did an additional constructability review of the coordination model. Examples of issues discovered in the constructability review were structural steel conflicts with the egress stairs, and the concern that the MEP systems were given insufficient space to allow for hangers and insulation.

Construction Phase

Bidding documents sent to the contractors included a special appendix that outlined roles and responsibilities of each trade for VDC. Upon award of the contract, the selected contractor would be responsible for developing a VDC implementation plan describing how they would set up and manage the project. The project was a publicly funded project, and often it is one of the requirements for public projects in the United States to accept the lowest bidding contractor. The contractor did not have prior experience with VDC. To comply with the contractual VDC requirements, the contractor hired a VDC consultant to set up the project.

The contractor's VDC consultant set up the construction coordination model, but did not use the project coordinate system already defined in the design intent model. By the time this error was discovered, the subcontractors had already set up their construction models for each trade

Figure 3.5.5 (overleaf)
Timeline with products and services

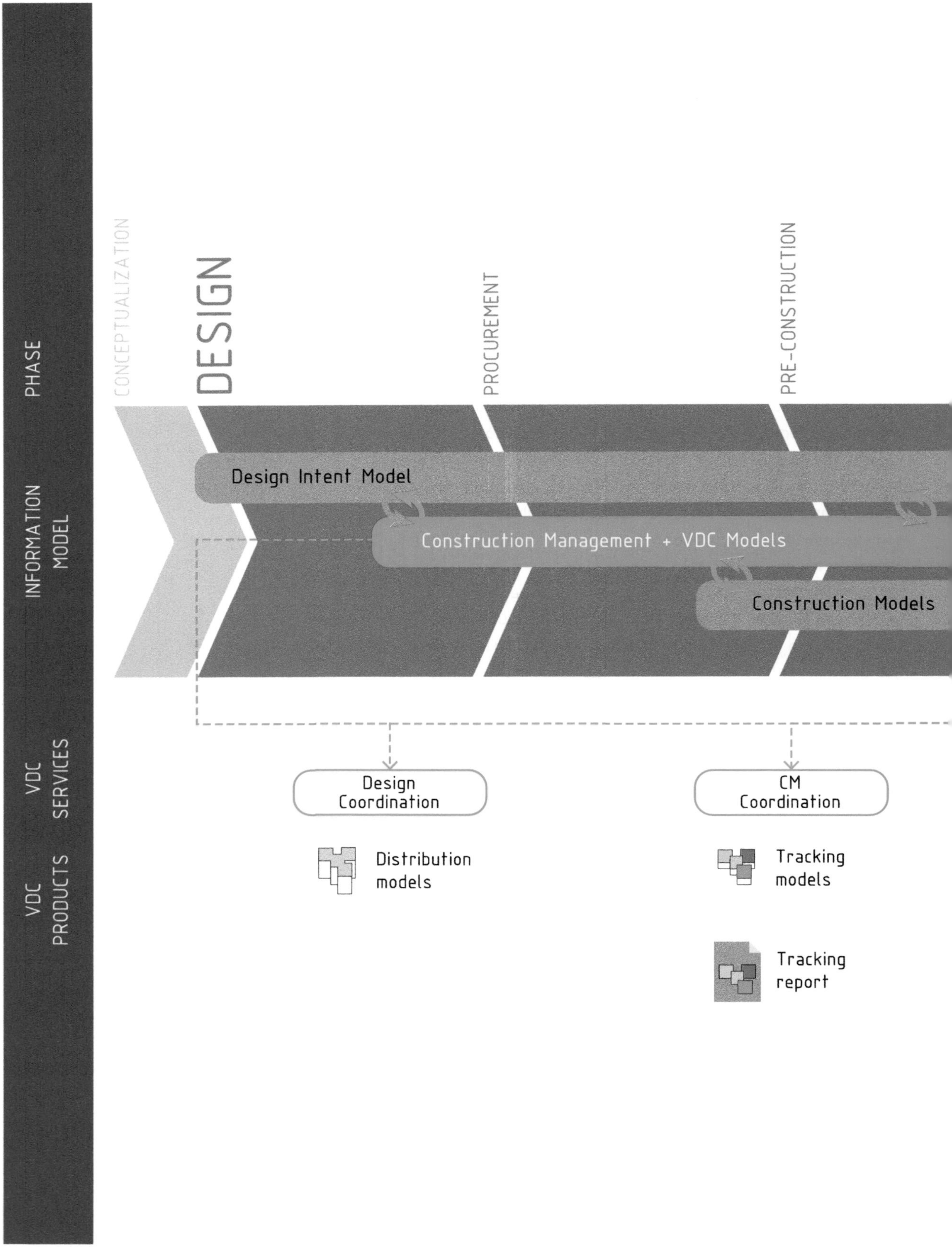

CONSTRUCTION

OPERATION

DECOMMISIONING

Experience
Simulation

Construction
Coordination

Tracking
Walk-throughs

Distribution
models

Clash
Reports

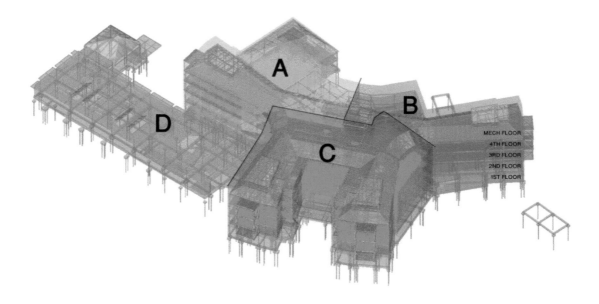

Figure 3.5.6
BMHR Adult Building area phasing
diagram

to the incorrect coordinate system. Consequently, it was difficult to set up automated week-by-week comparisons to ensure the construction model followed the intended design—a standard VDC deliverable. Once the coordinates were set, recreating the models in a new coordinate system would have been too expensive, so it was decided to keep the discrepancy between the models. The issue was not discovered in time to fix, because the contractor replaced their VDC consultant and hired an in-house construction VDC specialist. During this transition, the CM team did not have access to the models for verification and review.

The MEP systems were coordinated by the contractor in Navisworks. Subcontractor models from different VDC and CAD platforms were imported into Navisworks to create the construction coordination model, which was broken down by level and area. The areas corresponded to the sequence of construction, and all the systems in a given area were coordinated before moving on to the next area. The contractor developed a coordination schedule based on input from all parties involved to ensure that manufacturing and installation could commence in a timely fashion.

LiRo's VDC team was present for the initial months of construction coordination to ensure that the coordination and tracking processes were in place. One example of an issue the team mitigated was that the contractor did not enforce the agreed-upon naming conventions for files received from the subcontractors. When working in the coordination

VDC PRODUCTS

CM Coordination Model
An information model that holds the design intent and construction models of the structural, architectural, and systems disciplines. Managed by the LiRo virtual construction coordinator and used to track the contractor's coordination effort.

Construction Coordination Model
An information model that holds the construction trades models. This model was managed by the contractor.

Construction Model(s)
Models for each system which include all elements, equipment, fittings, etc., accurate in size, shape, location, quantity, and orientation with complete fabrication assembly and detailing info as required in LOD 400 (AIA-document E202). Each model is broken down by system, area, and level.

Distribution Model
The most recent Navisworks version of the CM coordination model, which is approved for distribution to the entire project team.

Tracking Model
A version of the design intent Revit model that was used throughout construction to record the weekly construction process. The geometric elements in the model were subdivided and assigned metadata relating to the inspection. This model supported weekly team meetings and serves as a record of the construction process.

VDC SERVICES

VDC Protocol Review
The VDC team reviewed and advised the development of the "BIM protocol" during pre-construction.

VDC Management
The VDC team worked with the CM team to oversee and manage VDC implementation during the pre-construction and construction phase.

Coordination Management
Most of the subcontractors were not familiar with VDC. The VDC team managed the implementation and setup of the MEP coordination process and subcontractors' information models.

Tracking
The VDC team tracked weekly construction progress in the information model. Visualizations from the updated information model were used in weekly progress meetings.

BMHR DESIGN INTENT COORDINATION STRUCTURE

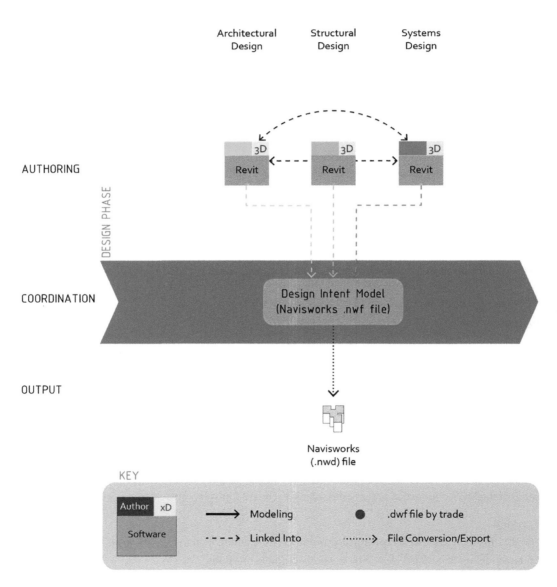

Figure 3.5.7A
Design intent coordination structure diagram

BMHR CONSTRUCTION COORDINATION STRUCTURE

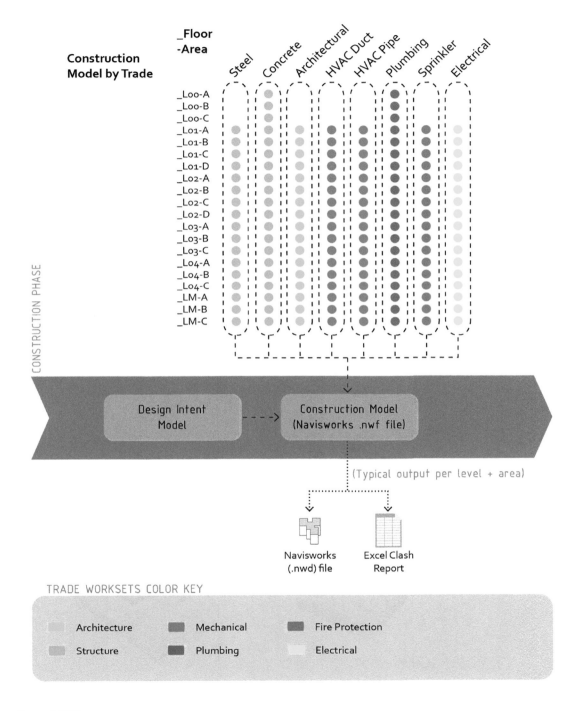

Figure 3.5.7B
Construction coordination structure diagram

BMHR DESIGN COORDINATION STRUCTURE

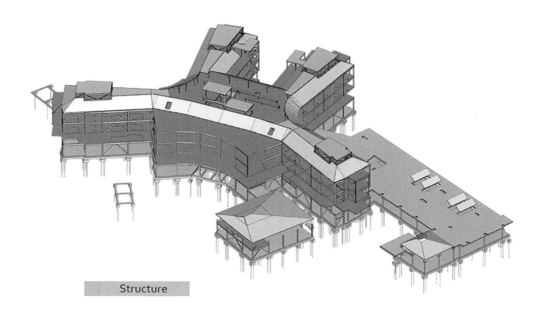

Structure

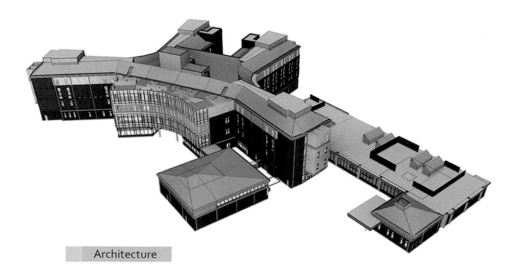

Architecture

Figure 3.5.8A
Model trade matrix—design intent coordination models

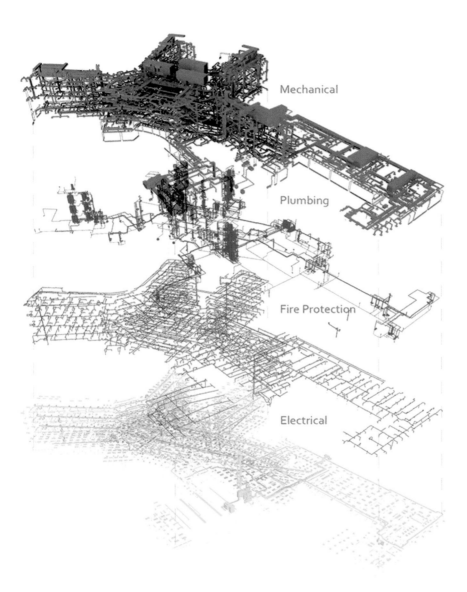

Mechanical

Plumbing

Fire Protection

Electrical

Systems

BMHR CONSTRUCTION COORDINATION STRUCTURE

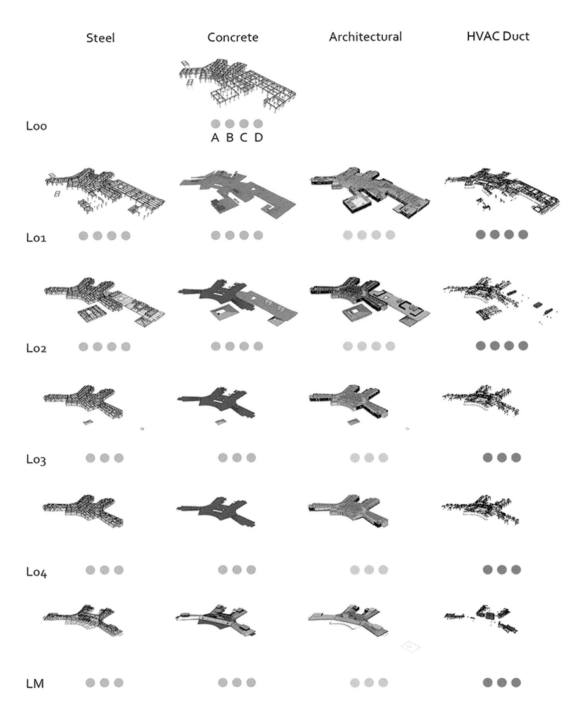

	Steel	Concrete	Architectural	HVAC Duct
Loo		● ● ● ● A B C D		
Lo1	● ● ● ●	● ● ● ●	● ● ● ●	● ● ● ●
Lo2	● ● ● ●	● ● ● ●	● ● ● ●	● ● ● ●
Lo3	● ● ●	● ● ●	● ● ●	● ● ●
Lo4	● ● ●	● ● ●	● ● ●	● ● ●
LM	● ● ●	● ● ●	● ● ●	● ● ●

Figure 3.5.8B
Model trade matrix—construction coordination structure models

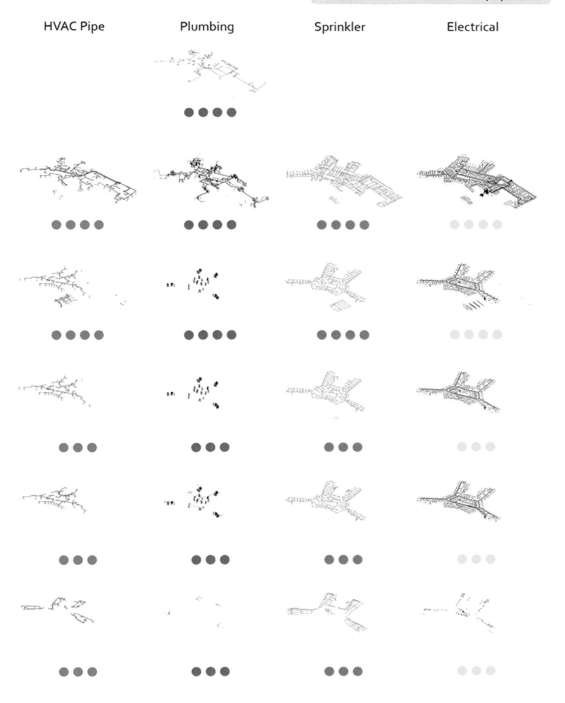

KEY ● .dwf file of trade and area A, B, C or D

| HVAC Pipe | Plumbing | Sprinkler | Electrical |

model, all the trade models need to maintain consistent naming, so that updates can be made easily by overwriting linked files. Due to the contractor's use of inconsistent file naming, the links in the construction coordination model had to be rebuilt for each clash cycle. Relinking prevented the team from tracking clashes as they were resolved each week, because the software treated each clash cycle as a brand new project instead of as an updated link to an existing project.

Because the LiRo VDC team was not directly supervising the subcontractors, it was difficult to make adjustments to the process, such as enforcing consistent file-naming standards. The larger lesson of the case study is that well-defined and enforced implementation plans and VDC protocols are critical to the success of VDC.

IN-DEPTH DESCRIPTION OF SELECTED VDC PRODUCTS

Tracking

Model-based tracking is an organized way of recording and reporting construction components as they are installed. Such tracking data are very useful, as it enables automated quantity takeoffs and gives a clear picture of ongoing resource allocation. With real-time data encapsulated in the tracking information model, it is possible to accurately measure if the project is ahead of or behind schedule. The data can be used to ascertain what portion of the project has been completed, and verify what work the contractors should be paid for, ensuring an accurate flow of money to keep the job moving smoothly.

Traditionally, contractor invoices have been paid through a largely manual process, in which the status of a construction item is recorded on paper on-site and quantities estimated in the field office. VDC uses a digitized tracking workflow, through which the tracking model provides a greater amount of detailed information to the team, including location, time, and quantity of each construction item.

At the time the project was estimated, effective information model-based cost-estimating software had not yet been developed. Had the appropriate software been available, real-time progress quantity information could have been extracted from the tracking model. The tracking model syncs to a consistent source to extract each item's costs, automating and refining the accuracy of some of the progress payments. However, as a field inspector ultimately carries out the data collection, the payment accuracy would still have hinged on the accuracy of the field inspections.

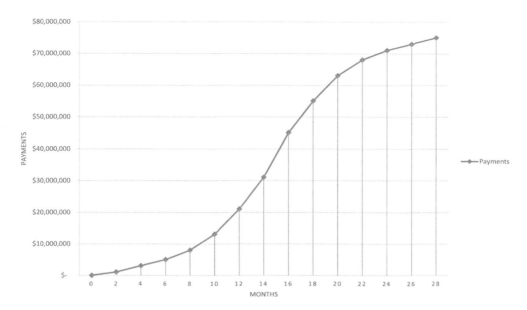

Figure 3.5.9
S-curve chart: cumulative payments on a typical project usually form an S-curve

The project team reviewed construction progress of the BMHR site in weekly progress meetings every Monday. Every Friday, an inspector recorded the progress on paper, and a VDC specialist input the data into the model. In order to better understand the project and the process of inspection, the VDC specialist started to conduct the inspections as well. In Monday meetings, the model was used to communicate the weekly progress of the project with a color-coding system that indicated the type of construction completed.

Tracking Workflow

The goal of tracking the project with VDC was to enable all of the tracking data to be processed and clearly visualized in one place, with minimal work required from the project team. Correctly implemented VDC tracking methodologies not only minimize time spent preparing tedious reports, but allows for nearly instantaneous project status updates. The moment the inspector syncs the tracking data, the team has access to that information. By planning all the tasks, applying correct configurations, and keeping things organized through well-executed VDC procedures, it is possible to understand and visualize the progress of the project as it is installed on-site.

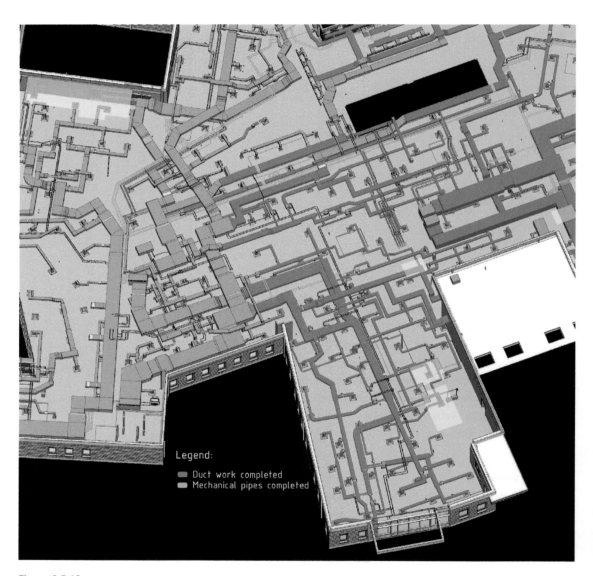

Figure 3.5.10
Color-coding of mechanical systems

BMHR TRACKING DIAGRAM

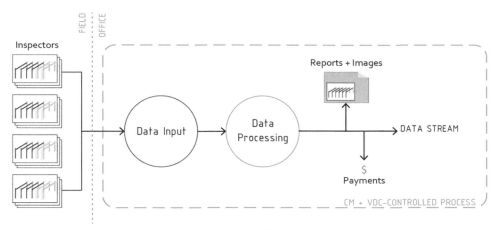

Data are input into the tracking model by a designated VDC
team member, using parameters we put in place.

Figure 3.5.11
Tracking diagram

In the Adult Building project, there were two parameters for each
tracked construction item: one checkbox to confirm that the item had
been completed, and a second indicating what week it was installed. After
progress data were collected in the field, usually on highlighting paper
plans, the information was taken to the field office and input into the
Tracking Model. All tracking data were housed in Revit, which performed
as a database. Tracking data was linked to the geometry, and input directly
through Revit. After all of the week's completed objects had been
checked, the week number could be assigned to all items in one batch.
The VDC team developed a Revit plugin to automatically update and
output the graphics based on the previous week's completions. The
output format was either PDF files for printouts, or 2-D CAD files or 3-D
DWF files that could be loaded into Navisworks and used live in
coordination meetings.

The tracking report's function is to clearly communicate the progress
of the project for contractors, construction managers, and all other
members of the team, so that they can clearly understand actual versus
planned progress. To communicate this information, the VDC team
generated a combination of color-coded plans and 3-D views of the
project, in conjunction with the planned schedule data. Staffing statistics
were also included in the report, so that productivity rates could be

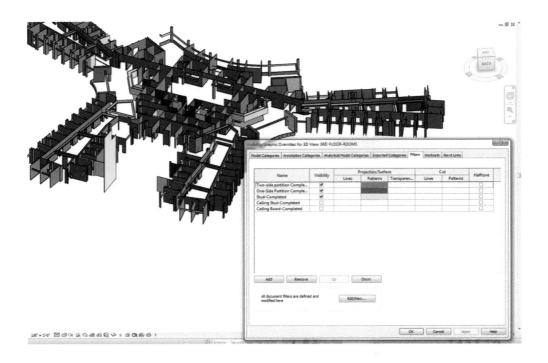

Figure 3.5.12
Screen shots of parameter checkboxes
and filters in Revit

calculated. For coordination meetings, 2-D plans function as a map that
clearly communicates location of progress, while 3-D model views help
convey geometrical relationships between components.

Detailed Tracking Animations

The client, located in Albany, NY, needed a way to keep up to date with
the construction processes without frequent trips to the Bronx site. The
data generated from weekly inspections were a good measure of overall
progress. By linking the model to the visualization engine Lumion, the VDC
team was able to provide the clients with weekly virtual walk-throughs of
the construction site in its most recent state of completion. This included
the different levels of finish on the partitions: stud framing, insulation,
gypsum finishes, and paint finishes. While it is difficult to measure the true
value of this tracking representation, it was done as an experiment to test
how far tracking visualizations can be taken. Something that might
previously have taken weeks to prepare could now be done in a day. Some
less technically inclined agency representatives benefited from a clearer
sense of the progress. If this methodology were to be linked to more
automated tracking workflows, it would be possible to access live progress
models from any device, and for any team member to review and
comment.

Figure 3.5.13
A sequence of images from a walk-through animation of the project's tracking progress. Animations, images, and tracking reports were shown at weekly construction progress meetings on-site, and shared with the client, who was based in another location

Figure 3.5.14
2-D plans and 3-D model views of construction
progress were created for tracking reports

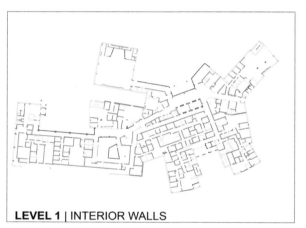

LEVEL 1 | INTERIOR WALLS

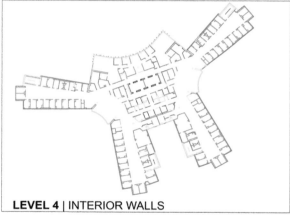

LEVEL 4 | INTERIOR WALLS

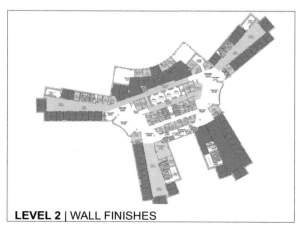

LEVEL 2 | WALL FINISHES

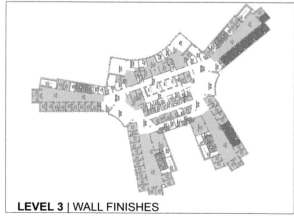

LEVEL 3 | WALL FINISHES

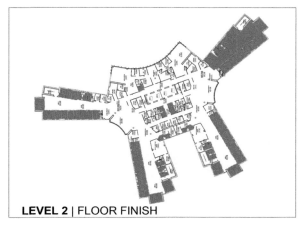

LEVEL 2 | FLOOR FINISH

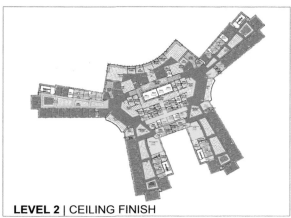

LEVEL 2 | CEILING FINISH

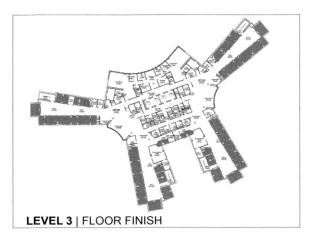

LEVEL 3 | FLOOR FINISH

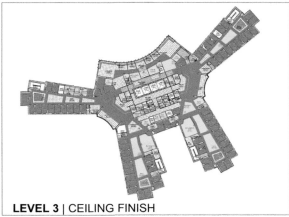

LEVEL 3 | CEILING FINISH

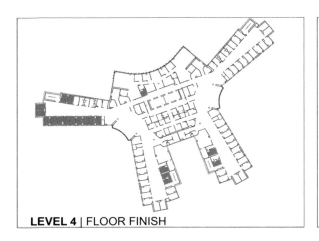

LEVEL 4 | FLOOR FINISH

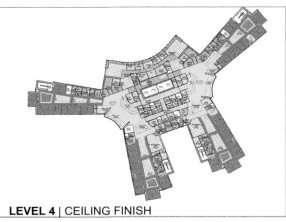

LEVEL 4 | CEILING FINISH

SOFTWARE PARAMETERS FOR TRACKING

Program Settings

In both Revit and Navisworks, certain settings are important for configuring an effective tracking workflow. In Revit they are Views, Sheets, Shared Parameters, Filters, Schedules, and View Templates. In Navisworks, the areas to configure are Model Configuration, Saved Viewpoints, Search Sets, and the Appearance Profiler.

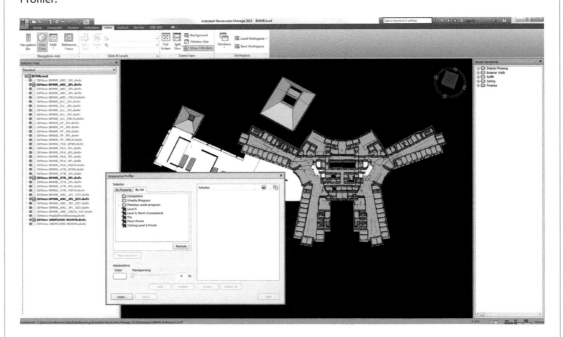

Figure 3.5.15
Navisworks screenshot with settings menus

A range of categories for building elements are tracked, and parameters are needed to reflect the diversity of tracking data. Some objects are tracked in a series of completion stages, and therefore need multiple parameters. In this case, there are two parameters for each tracking task: a simple checkbox for completed work, and the week number the element was completed.

Certain object types are broken up into smaller construction pieces for accurate sequencing. One example is slab pours, which are often done in pieces. The Part Tool in Revit is used to subdivide those components.

Items Tracked on BHMR Project

Exterior

- Walls
- Roof
- Glazing

Interior

- Partitions and soffits
    - Studs
    - One side gypsum board

- – Wiring
- – Closure of wall
- Ceilings
- Doors

Systems
- Mechanical
- Electrical
- Plumbing
- Fire protection

Furniture, Fixtures, and Equipment
- Furniture

- Fixtures
- Wall panels
- Window sills

Room-based Inspections
- Finishes
  - – Floor
  - – Walls
  - – Ceilings
  - – Doors
- Punch list items

Punch Listing

The punch listing stage is the final stage of the construction process, leading up to turnover of the facility to the owner. The purpose of the punch listing period is to record and report outstanding items or defects on areas deemed to be completed. As an alternative to the typical paper punch list procedures, the CM team used an iPad app called Newforma Punch List. Not all item issues relate directly to model objects (examples of those could be cleaning and patching); however, each punch list issue was assigned to a room and space, which were exported from the information model and used as the basis of the punch list. Punch list items were associated with a room and accompanied with a photo and written description of the issue. The items were assigned to the responsible party with a due date and automatically distributed among the members of the team. This digital workflow made it easy to log and report punch list statuses and track any outstanding items. Reminders were automatically sent out as the deadline of each item approached.

INTERVIEW WITH MARK SWANSON

Mark Swanson is the CM Project Manager for the LiRo Group on the Adult Behavioral Health Center of the Bronx Hospital Redevelopment project. Mr. Swanson was interviewed by Lennart Andersson at the site of the project on July 18, 2014.

What is your role on the project?

M. S.: I'm the CM Project Manager for the Adult Building.

Can you describe the project?

M. S.: The project is made up of six buildings: the Residential Village that consists of three buildings, the Children's Building, Central Services, and the Adult Behavior Building. LiRo is the construction manager of the Adult Behavioral Health Center, while Jacobs is the CM for the other buildings. The design started around 2010 with the Adult Building, which was the first building to break ground in 2012. It is a 180,000-square-foot ground-up new building—five storeys, 156 beds, steel construction on piles.

Why was VDC piloted on the Adult Building project?

M. S.: About five to six years ago, DASNY had a directive to start using 3-D technology for design and construction of projects with the goal of making things more efficient and saving money and time. Initially, when the project program was being defined in the early planning stages, there were meetings with architectural firms. It was decided that VDC would be used for the design of the project. There are three architectural firms on the project: STV, Architectural Resources, and Spector Group.

Did you have any prior experience with VDC? What was your perception of VDC?

M. S.: Not before this project. All I knew was that it involved 3-D models that you could visualize on the screen, but I didn't know how it would be used as a tool and incorporated into the construction process. I was fascinated with it and thought it would be something interesting to use on this project. I didn't know what to expect.

Have you been involved with the implementation of VDC on the Adult Building project from the start?

M. S.: Yes, we were involved from the very start. In the beginning, we took part in developing a document that came to be called Appendix O in the bid documents. It was essentially a guideline for all bidders and contractors on what their responsibilities would be. It also outlined the architect and CM responsibilities during construction. It was reviewed by countless numbers of people. The document was amended numerous times as everyone started gaining experience [with VDC] and developing a better understanding of the process and changes in the technology.

In your opinion, who benefits most from VDC?

M. S.: I would say the biggest benefit is that it actually forced the contractor and their subs to really study their layout of work in greater detail than they normally do using traditional coordination tools, such as Mylar or CAD layovers.

How did VDC diverge from your expectations?

M. S.: It's been interesting to watch it evolve and see how the various design consultants have designed their buildings using this technology, and the pros and cons of that. It's also been interesting to see how the entire contracting community has reacted to VDC during the bid phase. Because very simply stated: The bidding community was not on board with it, not informed or up to speed as far as what the technology was and how to use it. One of the biggest issues was cost; which software to use, how to train people, who would do the data entry, and so on.

All of that was brand new to the bidding community, when these jobs were all bid two and a half years ago. So there was a tremendous learning curve for the majority of the construction community, except for certain trades that didn't use Revit, but were already modeling things in 3-D using other types of software specific to their trade. Steel and duct work historically were modeled, as they tie models directly to the manufacturing equipment. Other than that, brick workers, carpenters, plumbers had never used this stuff before. So it's been very interesting to watch it evolve into what it is now.

How would you evaluate the failures of VDC on this project?

M. S.: I would say one of the negatives was that the designers designed the model to LOD 300, which omits important details as far as pipe or duct wraps or hangers. The contractor is responsible for this, and it's an area where issues often happen when fitting everything within a ceiling, which resulted in numerous cases where the ceilings had to be lowered to accommodate everything above ceiling height. That wasted time and resulted in some change orders during construction. Another negative was not in its [VDC] execution, but in the requirement to maintain an as-built model of the project on a weekly basis. It is unrealistic for a project of this magnitude to require the continuous updating of the as-built model. It should be updated on a regular basis as the work occurs.

Has VDC changed the way your team works?

M. S.: Yes, we went from traditional ways of doing things to modern ways of doing things in 3-D. The whole idea of doing MEP 3-D coordination instead of the traditional way with overlays with Mylars back and forth (which has been the standard for decades) is a completely different way of thinking, different way of approaching something. It changes your management style, as now you have to think outside the box. You are looking to use it as a tool to figure out problems, because you didn't have it before. Looking at change orders, looking at design issues. In the past, you had 2-D drawings to look at, and now you can actually go into the system to analyze the impact of changes, and how they relate to other assemblies in three dimensions. From a management standpoint, it opens up a lot of doors to solve problems in a way we were not able to do before.

If you started a new hospital project today, how would you do things differently?

M. S.: The architect would need to design to a level of detail that includes all dimensionally accurate elements. I would make sure the language in the specifications was very clear as far as the architect's management of the VDC process. We tried to create specifications and thought we had it pretty well detailed, but they were limited to what we knew at the time. They turned out in some cases to be a little bit unrealistic, and some language was weak in detail. Now, it's clear the contractor needs people that have the skill set to manage the VDC, input data, and maintain the documentation. They need to be able to train the subcontractors to get them on board. We need a more dedicated effort from the contractor compared to what we experienced on this project. Moving forward, there needs to be a better effort to utilize the technology for what it is meant to do. There has been a tendency, sometimes, to forget we have VDC technologies when we are in a hurry and pushing to get the job done, then we worry about VDC later on. But VDC needs to be more integrated into the everyday workflows. One thing we've started to do is use it as a tool to track construction progress, which has proven to be a very helpful tool. We will incorporate VDC from the beginning, from the foundation and up.

Overall, has VDC helped save money or time, considering that this was a pilot project for most of the people involved?

M. S.: That will be a tough thing to analyze. I would tell you, based on the learning curve from what the bidding community did not know up front, and considering the time it took to get things organized and work out all the bugs upfront with the contractor and its subs—there was time and money lost there. But I think, at the end of the day, it will level itself out. I am not sure it will be a money maker on this job, but I can see going into the next project with what we know now, and the lessons we learned from our experience on this project—I can see it becoming a true money maker on the next project, making things much more efficient and streamlined, because we know a lot more than we did in the beginning of this project. This project was really the guinea-pig utilizing VDC both in design and construction. Moving forward it will work a lot better.

Do you think the Facilities Managers and the end users will benefit from VDC?

M. S.: It depends on the client. If you have a client that is financially sound and has a lot of resources from a staffing standpoint, then they can actually utilize the technology. The client needs to have people on board that can be trained and the resources to use VDC for Facilities Management. State [publicly funded] projects tend to be short on staff; they don't have deep pockets. They tend to have old and antiquated technology that is behind the times, but that is what they can afford. When we use this great technology for construction and it has to be melded with old facilities software, it is a tremendous conflict as they don't match up. The FM and state organization is not going to say they're going to bring on a whole new team of people and revamp the whole system. At the end of the day, they will receive as-builts in VDC that will detail everything in the new project, but there will be a limited integration with their facilities management system.

What would be the most important thing to teach a Construction Manager about VDC?

M. S.: I think it should be a multifaceted training starting with topics like what is the technology and what does it do? Start by explaining the language, the acronyms. The language of VDC—it is a whole other language! I was not familiar with it until I was exposed to it. Begin with an upfront exposure of what VDC is and what it's made for, and then break it down from there. What are the different components, how do they function? Does the construction manager have hands-on experience to sit by the keyboard and actually work on the model? It would be nice. I have tried it. I have no idea what I am doing. We have some great equipment here. I have sat down and said "OK, lets push this button." It would be great to learn the basics, how to navigate a model. How to go into a model that someone else is actually maintaining. If I want to go into a model to look something up, like a detail, a column, a beam, or a floor plan. How do I do that?

Maybe it would be beneficial to have a hands-on, compact instruction book, like a cheat sheet on how to do things. We end up investing a lot of money in new equipment, but it does not get used to the extent that it could. For the next project, now that we have an understanding of how it works, some additional training would go a long way.

CONCLUSION

The BMHR project was one of the first projects in the New York State area where VDC processes were used all the way from design through construction. Many of the people involved with the project had never been exposed to these new concepts. It was, therefore, quite a learning experience for all parties. One of the major lessons learned was the importance of including hangers and insulation in the design intent MEP models, in order to truly be able to coordinate everything that goes on above a ceiling before the construction phase begins. Another lesson was that the facilities requirements need to be worked out with the owner before the contract is awarded, so all the relevant data can be added to the information models as the project progresses and be available for upload to the facilities management software at the end of construction.

In terms of major hurdles, when the project was started in 2010, VDC tools for cost estimating and 4-D scheduling were not sufficiently developed to be used effectively. The hardware was also not yet powerful enough to effectively drive the large, complex information models required for a project this size. This sometimes made it difficult to use the models for ongoing live coordination, as the hardware tended to behave sluggishly. It was really only during the last year of coordination that sufficient hardware was finally available so that the models could be utilized effectively for the project.

Based on experience gained in the Adult Building project, for future projects of similar scope, the design team should be required to model to the more detailed LOD 350 standard in order to perform meaningful coordination. The CM team should be brought on board at the beginning of the construction document phase, in order to use the model to generate meaningful cost estimates and 4-D schedules, which are powerful feedback instruments to ensure that the project is delivered with sufficient quality, on time, and on budget. Even though most public projects don't require pre-qualification, it would greatly help the project to ensure that the contractor has the capabilities to utilize VDC to its fullest potential. The total amount of Change orders amounted to about 5 percent of the contract sum, which is considered normal but not quite good enough for a fully implemented VDC-run project (it ought to be closer to 0 percent). However, as a pilot project and an educational experience, the VDC effort was a major success.

4 VDC Tools

Many tools for the AEC industry have not seen dramatic change for many years. CAD-based drafting is little more than a computerized drafting table, with a workflow designed to generate 2-D paper drawings as the final product. VDC tools have a fundamentally different goal when compared with CAD and drafting tools. VDC processes create full, detailed simulations of projects, whereas traditional methods create only partial representations of the project. CAD lacks the data component of BIM. The enormous amount of data generated and processed by VDC on a project requires specific tools for effective data management.

Laser scanning, photometrics, RFID tagging, and drone technology can feed high-resolution, accurate data into the information model, establishing automated relationships between the virtual and real. During construction, robotic total stations make it possible to guide and verify the layout of the assemblies directly from the construction information model, so that tracking and as-built information can be incorporated into workflows on an ongoing basis.

VDC is maturing from a desktop-centric ecosystem with limited numbers of large, expensive software licenses into a complex, interconnected ecosystem in which data need to flow between many smaller, cloud-based software applications and different types of hardware devices, like tablets and phones. Computer software and hardware are both increasing in capacity and speed at an exponential rate. Many VDC tools are used not only in the office but also in the field, where much of the innovation is taking place.

SOFTWARE

Software used for VDC functions can broadly be broken down into three categories: model authoring, analysis and collaboration, and project information software. Some software applications perform functions in multiple categories, but generally, each individual application is most useful for its specific functionality.

Innovative plugins and apps are further enhancing the versatility of information models. The emergence of an all-encompassing VDC ecosystem is dependent on the interoperability of the different platforms,

LiRo VDC Tools

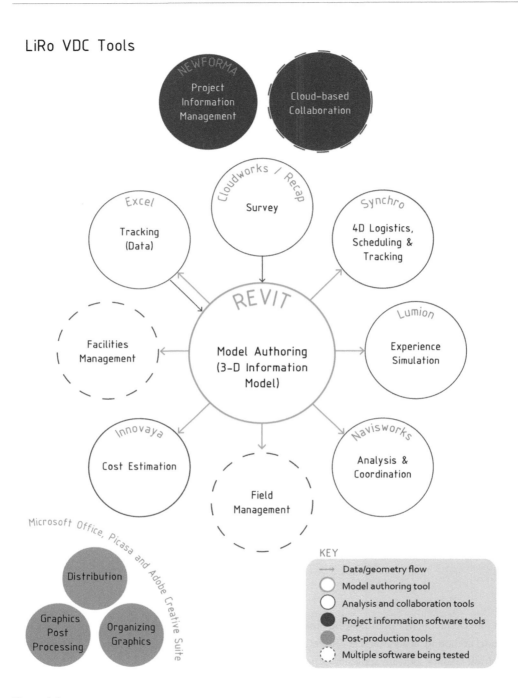

Figure 4.1
Diagram of software used by the LiRo VDC team

software applications, and file types. Further developments in this area will allow for more collaborative workflows.

Even though it is tempting to constantly use the latest version of a software, project requirements and compatibility issues in the software and hardware ecosystem sometimes force project users to continue using an older version. This is particularly true now, when programs are often not backward compatible, and extensive reliance on plugins, apps, and customizations requires these elements to be updated as well.

The VDC department is responsible for keeping up with new technological developments and testing new software for compatibility with current systems before upgrades are made to the larger organization's workflows. For larger organizations, hardware and software can be quite costly, as new versions need to be tested in order to ensure they do not conflict with the overall intranet of software used by the company. At LiRo, the VDC department runs its own network, in order to maintain the ability to constantly upgrade the software.

It is helpful for a cutting-edge VDC department to be involved with the software developers, in order to alpha and beta test the next generations of software. This enables workflows to line up with the latest technologies so that it is possible to internalize the new features when new software is released. Keeping systems simultaneously robust and flexible to continuous updates requires a clear vision and strategy.

When specifying VDC tools for the construction site office, it is crucial to include the appropriate hardware, software, and network connectivity so that VDC processes can become a seamless part of day-to-day fieldwork.

Modeling Software

Modeling software, also referred to as authoring software, is used to model geometry and much of the associated data. The most prominent authoring software programs are Autodesk Revit, Vectorworks, Tekla BIM, ArchiCAD, Catia, and Microstation BIM.

Analysis and Collaboration

Certain software is used to apply a range of specialized tools for analysis, collaboration, and data manipulation to the model created in the authoring software. This can include quantity takeoffs, project markup, scheduling, conflict resolution, code checking, flow analysis, energy analysis, or visualization. Some examples of this type of software are Rhino Grasshopper, Autodesk Navisworks and 360, Synchro, 3-D Max, Innovaya, and VEO. Some authoring software programs, such as Revit, have built-in analysis tools as well.

Project Information Management

Project information management systems link data from the model to distribution platforms, which record, organize, and track documentation, communication and decision processes. These programs are typically accessible through any type of device with a web browser. A couple of examples of those are Newforma, GTeams, Autodesk 360 Glue, and Field.

2-D VS. 3-D

That 3-D visualization is always preferable to 2-D is a common misconception. A 2-D plan or section function as a diagram, in which it is easy to locate certain things, as it includes rooms, grids, and levels. Seeing spatial relationships and taking measurements is often easier in 2-D. It can be difficult to understand where exactly one is in a 3-D model, and how an area relates to the whole. Especially when one is being zoomed into a

Figure 4.2
Two screens, one showing a 2-D plan and another showing a corresponding 3-D view

place inside the model and it's not clear which direction one is facing. In some cases, looking at a 3-D model can result in an information overload, and the benefit of having a simulated reality of the project is diminished. Time can easily be wasted in coordination meetings just searching for the location of a specific room in the model. The best method is to use 2-D plans as a map, and then turn to 3-D views to understand the specific conditions. A good analogy would be how Google maps use the 2-D map to describe a location, and street view to show specific context.

When using models, it is best to use two screens: one to display the model in 2-D, and the other to display 3-D and data to avoid having to switch back and forth between the two. Revit is one of the few software applications that handles both 2-D and 3-D quite well. Unfortunately, none of the coordination software available does this effectively, as it seems the myth of the total superiority of 3-D visualization still prevails.

APPS

Independently developed applications are emerging that attempt to augment the current workflow in the AEC industry. Technology start-ups are increasingly present in the construction space. The construction industry is ripe for disruption, and because much of the industry's work takes place outside the office, mobile apps are a logical next step. However, interfacing with construction drawings requires large tablets and instant computational and internet speeds in order to function effectively.

The greatest need for app development is in tools that can deliver information to the construction site as well as seamlessly collect data in the field. A few large companies have monopolized most of the construction industry's technological infrastructure, mainly Autodesk, whose Autodesk 360 Field is one of the ubiquitously adopted mobile applications. It does improve the existing workflow by connecting information models to data collected from the field using tablets and mobile devices, but it is primarily focused on equipment commissioning and inspection. Furthermore, Autodesk 360 Field lacks comprehensive data tools for tracking and does not have an effective interface for linking drawings and models. There is plenty of room in the industry for lightweight applications developed by independent companies.

SELECTED SOFTWARE: OVERVIEW

Modeling

- Revit Architecture—architectural and structural modeling
- Revit MEP—mechanical systems modeling
- Rhino—free-form and generative modeling
- Catia—advanced modeling to fabrication

Analysis

- Navisworks—constructability, visualization, clash detection, and sequencing
- Synchro—4-D scheduling visualization
- Innovaya—cost estimating
- 3-D Max—rendering and animation
- Lumion—experience simulation
- 360 Glue—cloud-based collaboration
- 360 Field—cloud-based field management
- Excel—tracking

Project Information Management

- Newforma—platform-neutral, information model-driven project information management

SOFTWARE: IN DETAIL

Autodesk Revit

Developer: Autodesk
Website: www.autodesk.com
Type: Information Model Authoring

The first version of Revit, released in 2000,[1] was the first true 3-D information-based building modeling tool, with its then revolutionary capacity to output traditional 2-D drawings from the 3-D model. Revit was acquired by Autodesk in 2002 and Revit gradually replaced Autodesk's failed attempt at BIM software, AutoCAD Architectural desktop application. Since then, Revit has grown into one of the leading information model authoring tools. Initially, Revit only supported architectural and limited structural modeling. In 2005, Revit Structure was released followed by Revit MEP in 2006. Since 2013 all three

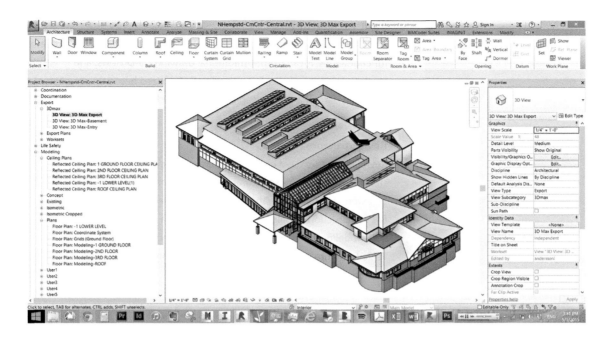

Figure 4.3
Screenshot of Revit software

disciplines have been bundled into one product. Revit has transitioned
from a tool focused primarily on the design disciplines, to one that is
now increasingly used during the construction phase by construction
managers, contractors, and subcontractors. Revit is still installed and runs
on PCs; however, multiple cloud-based tools have been added. The
number of specialized plugins is constantly growing as the software's API
continues to be developed.

Revit's greatest strength also has a drawback: the program now has
so much functionality built into it that professionals face quite a steep
learning curve. But, Revit is unparalleled in its capacity to display 2-D
drawings generated out of a 3-D model. Almost all other BIM
applications lack this key function—a way to clearly view the model
through traditional methods such as plan and section. Revit is also a
powerful database; one shortcoming is that developers have not yet
added effective tools for searching and organizing data aside from the
scheduling functions, which have not been significantly altered since
early versions of the program.

Revit Architecture

Revit Architecture consists of all the architectural components and relating functionalities. It can output traditional drawings, 3-D views, and renderings. Elements can be tagged and visualized based on conditional data. The data of a project can be output on schedules.

Revit Structure

Revit Structure is used by the structural engineer to document the structural design. The structural calculations are often still performed by external applications, then imported into Revit for documentation. For construction detailing, there seems to be a lack of support for LOD 400 modeling, so often this is done in other modeling software, such as Tekla Structure, and subsequently imported into Revit through the IFC format.

Revit MEP

Working with BIM, it becomes clear how complex and time consuming the MEP portion of both design and construction is. Revit MEP can be used to design mechanical, electrical, plumbing, and fire protection systems. It includes tools to calculate heating and cooling loads, plumbing systems, and electrical loads. Since much of traditional MEP documentation is quite diagrammatic, MEP trades have been somewhat reluctant to adopt Revit, as some of its drawings do not display details as clearly as traditional drawings.

Revit Plugins

Revit has a whole range of plugins, especially since the development of Revit API. The following are some of the most commonly used at LiRo VDC:

- Imaginit tools—data link to Excel tools
- Scan to BIM—plug-in for modeling of objects from point clouds
- Newforma—documenting issues in Revit for distribution through Newforma
- COBie Extensions—enable export of data for facilities management
- Point Layout—enables point layout for field
- SysQue Hanger plugin—creates hangers for MEP systems

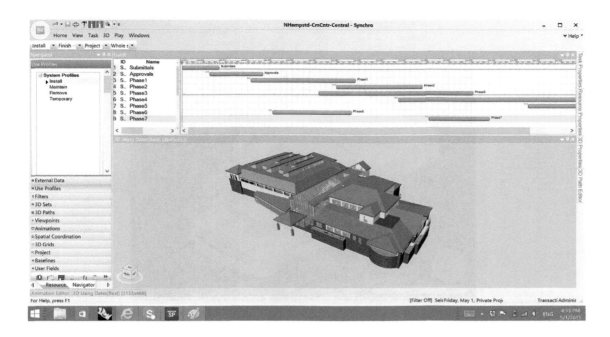

Synchro

Developer:	Synchro LTD
Website:	www.synchroltd.com
Type:	4-D and Scheduling
Requires:	3-D geometry from Revit or Rhino; schedule can be built in Synchro or imported from scheduling software such as Primavera, Asta, or Microsoft Project
Key functionality:	Links 3-D model to the construction schedule (time), essential for 4-D visualization. Also used to visualize construction progress.
Alternative:	Navisworks Manage

Figure 4.4
Screenshot of Synchro software

Synchro connects an information model to the time component of a schedule. It accepts Primavera-generated .XER files and 3-D geometry generated in authoring software, such as Revit. Synchro generates a 4-D model by connecting the schedule and the 3-D geometry. At its core, Synchro is a full-feature scheduling platform that internalizes all of the functionalities available in other scheduling software, and it can be used as a main scheduling tool during the planning and monitoring phase. Synchro makes it possible to run comparative scenarios between multiple schedules, and color-code a model based on multiple parameters like time, activity, or stakeholder.

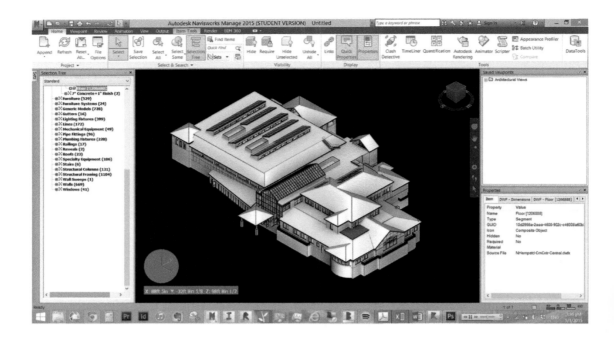

Figure 4.5
Screenshot of Navisworks software

Navisworks

Developer:	Autodesk
Website:	www.autodesk.com
Type:	Coordination
Phase:	Analysis and Collaboration
Requires:	3-D geometry from Revit
Key functionality:	3-D review, clash detection, sequencing animation, schedule visualization, and quantity takeoff
Alternatives:	360 Glue, Synchro, Innovaya, 3ds Max

Navisworks includes a range of functionality for reviewing and
coordinating projects. Its primary benefit is its compatibility with a wide
range of model formats. It makes managing large, detailed projects
easier. Views can be saved and organized for effective review. Sections
can easily be cut and dynamic color-coding can be applied and saved
within views. The clash detection engine is highly developed, so multiple
clash sets can be run and reviewed. It is possible to run fairly complex
sequencing animation in order to understand and communicate how
parts of buildings actually need to be assembled. Schedules can be
linked to a model and visualized, but actually working with schedules in
Navisworks is inefficient. Synchro remains the preferred software in this
regard. Navisworks' quantity takeoff tool, introduced in 2013, is

becoming fairly advanced. However, it has yet to be linked to a cost database for automated cost estimating as in Innovaya, a more specialized tool for this purpose.

Autodesk 360 Glue

Developer: Autodesk
Website: www.autodesk.com
Type: Coordination
Phase: Design and Construction
Requires: Model synched from authoring software
Key functionality: Cloud-based, 3-D visualization and collaboration of
                       information models
Alternative: Navisworks

In simplest terms, Autodesk 360 Glue is in some ways a cloud-based version of Navisworks. Models can be uploaded and synched from a range of software, creating a merged model that can be used for collaboration. Views can be saved and accessed easily, so sharing them with existing team members and bringing new members up to speed is simple. It's also easy to cut sections, and the software includes annotation tools like redlines and dimensioning. The models can also be viewed on mobile

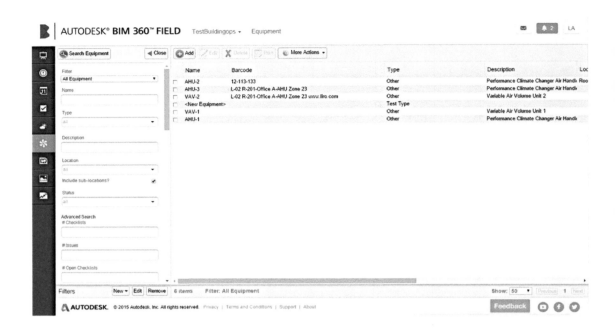

Figure 4.7
Screenshot of Autodesk 360 Field
software

devices like iPads or iPhones. Components can be added to equipment lists, which can then be pushed to 360 Field in order to generate linked inspections lists. The software's primary weakness is that it cannot view 2-D drawings in conjunction with 3-D. There is also no way to manipulate the data generated from the Revit model.

Autodesk 360 Field

Developer:	Autodesk
Website:	www.autodesk.com
Type:	Field management tool
Phase:	Construction
Requires:	Checklists can be created, imported from Excel or a model imported from 360 Glue
Alternative:	Excel, Newforma field tools

360 Field is a powerful field management tool with capabilities for inspection and commissioning activities. It links the field to the office and offers reporting tools, though it lacks a way to perform continuous tracking of components other than equipment. It does not offer a way to work directly from the model/2-D view, but is rather operated through a series of forms.

Lumion

Figure 4.8
Screenshot of Lumion software

Developer:	Act-3-D B.V.
Website:	www.lumion3d.com
Type:	Architectural visualization and experience simulation
Phase:	Design
Requires:	Model geometry synched from authoring software
Key functionality:	Realistic visualizations and walk-throughs with fast turnaround time
Alternatives:	3ds Max, Revit, Cinema 4D, Navisworks

Based in the Netherlands, Act-3-D was one of the pioneering forces behind real-time 3-D software when 3-D acceleration hardware came about. After several years of experience in simulation, training, architecture, TV, and film, Act-3-D decided to focus on architectural visualization.

Lumion creates renderings and walk-through animations. People, water, trees, and special effects like fire and smoke can be quickly rendered. Its reliance on a gaming engine gives Lumion a phenomenal frame rate, creating a feeling of live interaction. The product's near-instant renderer offers fast turnaround time, while still conveying an acceptable degree of realism as to what a space or project will look like. Lumion is much faster than its architectural visualization counterparts like 3ds Max and Maxwell render; however, it falls short of achieving their ultra-real quality. Not only are its rendering speeds fast, but its interface is also intuitive, making Lumion a good solution for anyone without professional rendering skills who wants to output a visualization quickly and cost-effectively. Lumion is not only useful for producing walk-throughs and renderings of the finished space, it can also be used to convey progress because of how quickly it exports animation. For example, producing frequent or iterative construction progress animations requires minimal effort, especially if a tracking model is already being updated for data and quantity takeoff purposes.

Innovaya

Developer:	Innovaya
Website:	www.innovaya.com
Type:	Quantity takeoff and cost estimating
Phase:	Design and Pre-construction
Requires:	Geometry from Revit or other model authoring software
Key functionality:	Quantity takeoff, cost estimating
Alternative:	Navisworks QTO

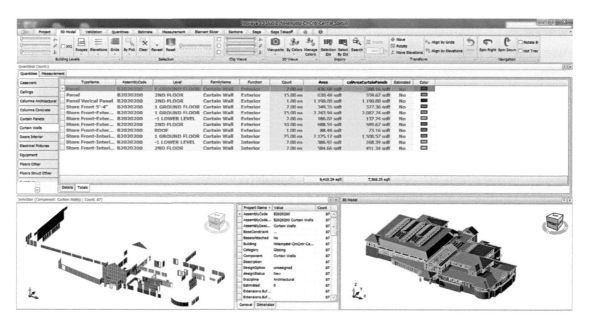

Figure 4.9
Screenshot of Autodesk Innovaya software

Innovaya is a quantity takeoff software that links to a cost database in order to create model-based cost estimates. It can apply automated calculations to elements from the information model in order to project accurate quantities not given straight out of a modeling authoring software. It has a model checking functionality that helps the cost estimator find problems with the model so that it can be used to extract accurate quantities. The schedule interface is linked to the model so that the quantities taken off can be visualized.

A Note on Databases

In order to harvest the power of data, it needs to be properly collected and stored. The ability to visualize data allows building professionals to plan and execute operations in more informed and efficient ways.

Revit acts as a database in the VDC workflow. Due to extensive knowledge with the program, the LiRo VDC team is able to "bend" standard features of Revit to fit desired workflow. The VDC team uses custom-designed Revit Add-In scripts to facilitate data exchange between Excel files, containing collected data, and Revit, where data is stored in the form of parameters attached to geometrical elements.

HARDWARE

LiRo VDC utilizes cutting-edge hardware optimized to run the various information modeling software. All workstations have dual monitors and enough processing power to run multiple BIM applications at once. Recently, the department has started to assemble its own hardware, in order to be able to continuously upgrade components as needed.

When using a software like Revit on large projects with multiple users, it is essential to have a fast file server with a network running at least one gigabit per second. A very fast wifi network is necessary for mobile use and meetings. The file server is equipped with two different drives: one ultra-fast, with wide bandwidth for ongoing project work, the other more normal, housing all the supporting files as well as inactive projects. Everything is backed up daily, locally as well as over the internet.

All machines have solid state drives, as they are faster and more reliable. A fast internet connection is essential in order to run coordination. LiRo built one high-powered workstation that can be accessed remotely, so all the software and models can be used from even a low-powered machine, which is particularly useful for coordination and meetings.

All members of LiRo's VDC team have administrative rights on their machines, in order to be able to continuously test different software.

This requires a certain level of trust and that everyone on the team have a level of understanding of how the system works.

For coordination meetings, two laptops are used with dual screens, so if one computer encounters problems, the coordination directly moves over to the other. A combination of laptops, tablets, and smartphones is used for ongoing project data collection and communication.

There is certain hardware that the VDC department interfaces with but does not necessarily control such as 3-D printers, laser scanners, drones, total layout stations, and other field equipment. These are, however, part of the ecosystem, and in order to perform effective VDC management, it is essential to have a good understanding of these technologies.

In order to stay current with the latest technology, it is best to have control of both the software and hardware, or at the very least develop a good relationship with the IT department.

NOTE

1   Arkin, Gregory, "The History of Revit—The Future of Design." Web log post.BIM Builder.com /Revit 3-D: The History of Revit-The Future of Design. n.p., October 7, 2007. Web. November 2, 2014.

5 Reference Documents

The success of VDC depends highly on organized collaboration. For a VDC practice to flourish, the VDC approach must be understood and agreed upon by all stakeholders on a project. The VDC team acts as a managing entity, ensuring correct integration and implementation processes. It is important for each trade to maintain strict adherence to standards provided by the VDC team while developing its model, in order to ensure uniform geometry and data formatting across all models. The project team must adopt hardware, software, and data-flow distribution systems at the beginning of the project in order to facilitate a seamlessly integrated workflow.

It is critical to employ well-defined standards to exchange, sort, and work collaboratively with data generated by the VDC process and stored in the information model. This chapter provides an overview of the LiRo Revit template, Navisworks and Synchro set-up and workflows, a sample VDC specification, a sample VDC implementation plan, and an overview of common classification systems used in construction and standards specific to information models.

LiRo VDC has developed a Revit template for model authoring that simplifies the set-up of projects and communication within the VDC ecosystem. This template is optimized for usage by the architect, the engineer, and the construction manager. Beyond the Revit template, LiRo VDC employs defined set-ups for clash detection and coordination in Navisworks and 4-D scheduling in Synchro.

Revit, Navisworks, and Synchro integrate with traditional building standards such as Uniformat, LOD 2013, and Work Breakdown Structures. Additional standards are used specifically for VDC. Some are still in development, such as the National BIM Standard and the IFC file format. To run a project effectively, the required standards must be clearly referred to in the specification documents.

A typical VDC specification document is included in this chapter for a project that integrates VDC throughout design, construction, and operations. The specification outlines the project goal and defines all the roles and responsibilities. To ensure that a project will be run according to the VDC specification, responsible parties for work outlined in the specification must submit a VDC implementation plan. A sample implementation plan is included in section 5.3.

5.1 Software Templates and Workflows

Fortune favors the prepared mind.

Louis Pasteur

Software set-up and workflows are crucial to the success of a VDC project. VDC software is powerful, but using the tools to their full advantage requires an experienced professional to create clear protocols and workflows. Correctly configured software templates give information modelers a head start when commencing a new project. Templates allow for knowledge about the technical aspects of modeling to be developed, transferred, and standardized across various projects, enabling better collaboration between members of a VDC team.

A well-developed template saves time during project set-up and deployment, and facilitates adherence to project standards. A good template provides a balance between extensive built-in standards, and simplicity and flexibility. As VDC encompasses all phases of a project, from design to construction and operations, clear, well-developed standards and workflows are essential for continual development of the information model. This section is an introduction to the templates and workflows created and used by the LiRo VDC team, which has made a considerable effort to analyze the needs of the design and construction process. The section presumes a certain level of familiarity with different aspects of VDC software. Please refer to the specific software guides for detailed information regarding the functionality discussed.

All the template files can be downloaded at www.buildingvirtual.net/templates

Videos that further explain the functionality of the templates and associated workflows are also available.

REVIT TEMPLATE ORGANIZATION

The LiRo VDC Revit Template has been gradually evolving over a number of years. It is optimized to cover both architectural design and construction management. The origins of the template, like the origins of Revit, are rooted in the architectural design phase. However, it has evolved with the experience of the VDC team and now contains standard set-up for both design and construction phases including systems (MEP) and construction management. The decision to create one master template for architecture, MEP, and construction management, instead of three separate more specialized templates, is meant to facilitate seamless collaboration and ensure cross-departmental standards within the LiRo Group. Annual software updates are also easier to perform once instead of three times.

The LiRo VDC template contains pre-formatted custom views based on the interests of different modelers at different stages in the building process. It is designed to expose the functionalities of Revit based on the interest of the modeler. For example, the architectural modeler is not necessarily interested in the systems and construction management views or sheets; therefore these views can be hidden with the aid of a pre-formatted filter.

The current template has grown to about 30 megabytes and can be easily managed with a standard, contemporary computer. The effective set-up of the template facilitates its ease of use. For example, the total number of views contained in the VDC template is around 2,000. Clear organization of the views in the project browser makes it easy to navigate the tree and find needed views. The VDC Revit template is designed based on the typology of a 40-storey building; therefore it contains multiple types of plans and views to efficiently model a large-scale project. The default typology of a multistorey building was chosen for the template because deleting views is much faster than creating them.
If the project contains only two floors, the set-up for the 38 unnecessary floors can quickly be deleted from the file. A typical drawing set for each trade can be generated from the model, as the title block and number standard are contained in the template. Project initiation sheets, with checklists that ensure a proper configuration at the start of a project, are also embedded in the template.

View templates are assigned to each view, and settings are locked down depending on use. Some views, which are designed to be printed out, have visual and line-weight settings configured specifically for printing, while other views are designed to facilitate an easier modeling process. A whole range of views assist in the quality control of the project.

The Project Browser
The project browser allows access to any view or sheet in the project. The default project browser shows every single view and sheet of the project in one long list sorted by type. Consequently, all plans or 3-D views are organized in a single list, which is acceptable for a smaller project, but for a large project, the flat organization of the project tree makes it difficult to manage the views and quickly switch between them. On a large project, the browser is easier to navigate if the views are organized by area or trade, rather than by type of view.

Figure 5.1.1
Project browser

Figure 5.1.2
Project browser organization

Views

In the VDC template, the project browser uses custom parameters in order to organize multiple types of views based on trade, view function, and type. Each view is defined by these three parameters, enabling the browser to be filtered by trade so that only the important views pertaining to the user are shown. View templates automatically organize any new view that is created.

The trade filters are as follows:

- Default—all trades organized by view type and then by trade
- All—shows all views, unsorted
- Architecture—only displays views used by architects
- Construction Management—only displays views by CM
- Services (MEP)—only displays views used by MEP modelers
- Services Drafting—only displays 2-D views by MEP engineers
- Structure—only displays views used for structural work

Depending on its use, each view is categorized by view type. These are as follows:

• 1. Modeling	Especially used for modeling
• 2. Documentation	Part of a drawing set
• 3. Presentation	Used for presentations
• 4. Export	Exported to other software
• 5. Analysis	Views used for analysis purposes
• 6. Project Controls	Views that help with set-up and management of a project

There is a third parameter in the project browser, the view subcategory, that helps to sort each view type. These are as follows:

Modeling

- Ceiling Plans
- Concept
- Demolition
- Existing
- Isometric
- Plans
- User-specific Views
- Worksets

Documentation

- Building Elevations
- Building Sections
- Ceiling Plans
- Demolition Plans
- Details
- Finish Plans
- Floor Plans
- Furniture Plans
- Isometric
- Life Safety
- Location Plans
- Notes and Symbols
- Perspectives
- Site Plan
- Slab Plans
- Wall Sections

Presentation

- Gross Areas
- Isometric
- Occupancy

Export Views

There are also views set up for exporting geometry to other VDC software. The views are mostly organized per workset and phase. These views are as follows:

By Project

- Demolition
- Existing to Remain
- New

By Level

By Specialty

- Visualization
- Cost Estimating
- 4-D Scheduling

Analysis

Analysis views are set up to help visualize certain specific requirements:

Architecture

- Insulation
- Open Area
- Room—Occupancy
- Sun Study

Structure

- Load

Services

- CFD
- Energy
- Heating and Cooling
- Lighting
- Pressure

Legends

Legends include lists such as abbreviations, symbols, partition types, and other legends. It is a good reference to help visualize the components loaded into the project (Figure 5.1.3)

Schedules

A properly developed information model contains a large number of schedules in order to access the data contained in the model. Unfortunately, the Revit project browser is not sophisticated at organizing and managing Revit Schedules. As the list of schedules grows in number,

Mechanical Components

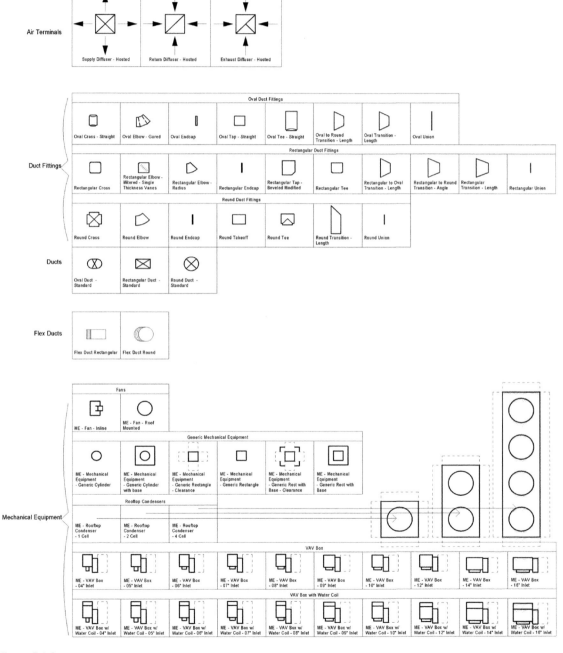

Figure 5.1.3
Mechanical components legend

searching for a given schedule becomes more difficult. To ease navigation of the schedules, the LiRo VDC team developed a naming convention to identify the use and type of schedule. Additionally, the schedules are placed on special scheduling sheets to further organize them by trade.

The schedule naming convention consists of a three-letter abbreviation at the beginning of the schedule, which facilitates sequential sorting in the schedule/quantity field in Revit. The following nomenclature is used:

- ARC—Architectural: schedules used on the architectural drawings
- PC—Project Control: schedules that help with model management
- QT—Quantity: schedules used for quantity information
- SRV—Services: MEP schedules
- CM-TRK—Construction Management Tracking: schedules used in construction management

Drawing Sheets

Revit views can be placed on sheets to create a drawing set. One can specify the trade of the sheet by right clicking on the sheet symbol in the tree and selecting the trade. To show all the sheets, simply select the default setting. In the VDC template, drawings sets are organized as follows:

Architecture
- Cover
- 0 Demolition
- 1 Construction Plans
- 1 Slab Plans
- 2 Exterior Elevations
- 3 Building Sections
- 4 Ceiling Plans
- 5 Enlarged Plans
- 6 Wall Sections
- 7 Vertical Transportation
- 8 Miscellaneous Details
- 9 Schedules
- Life Safety

Electrical
- Electrical Lighting
- Electrical Power
- Fire Alarm
- Nurse Call

Mechanical
- Fire Protection
- HVAC
- Medical Gas
- Piping
- Plumbing

Presentations
- Concept
- Gross Area Plans
- Occupancy Plans
- Programming
- Sun Study

Project Controls
A unique set of sheets for quality control that specifies the template version, information, and checklists of model requirements.
- Cover
- A: Project Management—template version, project Information, project location, design checklists, LEED checklist, project file organization, keynotes, and project scope
- B: Project Views—includes browser organization, schedules of the different view types
- C: Graphics and Materials—includes a list of the different lines and fill types
- D: Project Components—filtered lists of components missing assembly codes, component lists, and detailed door data
- E: Quantity Takeoff—quantitative lists

Sketches
- Architectural Design—sheets for sketches issued by the architect during the design stage
- MEP—sheets for sketches issued by the MEP engineers
- Architectural Construction—sheets for sketches issued by the architect during the construction phase
- CM—sketches made by the construction manager

Structural
- Plans

Families

All commonly used components are loaded into the project as Revit families. As with schedule organization, by default, project families are listed in one long row in the project tree based on component type. The project tree lacks the ability to organize families by trade or function. Therefore, a naming convention is used to identify which families are used by which trade. All annotations, title blocks, symbols, and tags are incorporated into the template and bundled under the annotation symbols category.

Only the most commonly used families are loaded in the template, modeled to a low level of development, as they tend to vary in detail from project to project.

The list below is sorted and based on type and use.

- Annotation Symbols
  - o   AN—built-in markers, such as section tags and view titles
  - o   SY—symbols
  - o   TB—title blocks
  - o   TG—tags, such as door or wall tags

- Detail Items
  - o   DT—architectural details
  - o   M—mechanical details

- Patterns
- Profiles
- Components
  - o   Architectural
    - – Casework
    - – Ceilings
    - – Curtain Panels
    - – Curtain Systems
    - – Curtain Wall Mullions
    - – Floors
    - – Generic Models
    - – Lighting Fixtures
    - – Mass
    - – Plumbing Fixtures
    - – Railings
    - – Ramps
    - – Roofs
    - – Site

- Specialty Equipment
- Stairs
- Walls

o Mechanical
- Duct Fittings
- Duct Insulation
- Duct Systems
- Ducts
- Mechanical Equipment
- Flex ducts

o Electrical
- Cable Tray Fittings
- Cable Trays
- Communication Devices
- Electrical Equipment
- Electrical Fixtures

o Plumbing
- Plumbing Fixtures

o Fire Protection
- Sprinklers

o All Services
- Flex Pipes
- Pipe Fittings
- Pipes

Groups

There are two types of groups in Revit: detail groups and model groups. Detail groups are only shown in unique 2-D views and model groups are bundled geometry and appear in both 2-D and 3-D views. Detail groups are an effective way to manage drafting configurations where 2-D lines, patterns, and detail components can be grouped and repeated throughout a project. Model groups should be used with caution. If components have associations with other objects or references outside the group, errors tend to crop up as the group tries to keep itself consistent. Only a few groups are incorporated into the template, as groups tend to be project specific.

Linked Files

Revit files can be linked to each other, for reference or for the purpose of constructing a master model. Linked files are managed by accessing a separate menu containing a list of all linked files. Linked models can be overlaid or attached. Files linked in for reference purposes only should be overlaid. If the linked file needs to be included when the host file is itself linked, then it needs to be attached.

View Templates

Using view templates ensures that the graphical settings for views are kept consistent. View templates also have parameters which can be used to automatically sort the views. The view template categories in the Revit template file are as follows.

- 3-D views
  - o Isometric
  - o Perspective—1 course
  - o Perspective—2 medium
  - o Perspective—3 fine
  - o Sectional Perspective
  - o X—fast
  - o X—isometric massing

- Ceiling Plans
  - o Documentation
    - – Architecture
    - – Services
  - o Modeling
    - – Architecture
    - – Services
      - -- All
      - -- Communication
      - -- Electrical
      - -- Fire Protection
      - -- Mechanical
      - -- Plumbing
  - o View Range—ceiling
- Elevations, Sections, Detail Views
- Floor Plans
- Schedules

View Filters

View filters in Revit are settings that can be applied to views to achieve different graphical results based on conditions. In the template, they are used to apply color to worksets and systems and subsystems, illustrate assembly code settings, identify insulation type, and visualize construction tracking. The filters can be applied to view templates, so when a new view is created the filter is automatically added.

Shared Parameters

Parameters contain data associated with Revit geometry. A parameter can be a dimension, text, number, or material, among other things. Parameters can be used to "drive" characteristics of the geometry, such as the length or width of a wall. Some parameters are built into Revit, but it is the program's ability to create new, custom parameters that makes it so powerful. The challenge with custom parameters is to keep them organized and eliminate redundancy. To schedule or export data held by parameters, the parameters must have consistent names. Shared parameters can be used across Revit files and projects in order to standardize them. LiRo VDC utilizes a shared parameter file, which has grown considerably over the years. The parameters are organized in groups based on their usage.

Parameter Groups

- General—parameters that are used most
- Clearance
- Codes and Data
- Construction Management
- Construction Tracking
- Family
- Facilities Management
- Materials and Finishes
- Misc.
- Views and Sheets

Tags

Tags are a powerful way of "reading" data from objects. Any component can be tagged. Tags can be made for specific components, or they can be multi-category and display data from many different types of components.

Figure 5.1.4
Shared parameters

Worksets

Worksets were initially introduced in Revit to enable multiple members of a modeling team to work on a project file simultaneously. Each modeler "owned" a workset which gave them exclusive rights to the objects contained in the workset. This prevented multiple users from manipulating the same object at once, which could cause conflicts or errors. Current versions of Revit track all elements and users within the software, and worksets are no longer as necessary for their original function. Worksets are still an excellent tool for model organization, but for an effective workflow, the number of worksets should be kept to a minimum.

The worksets used in the VDC template are named and the trades are bundled together:

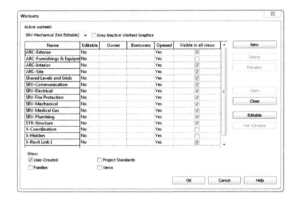

Figure 5.1.5
Worksets

- ARC (Architecture)
  - Exterior
  - Interior
  - Finishes

- SRV (Services—MEP)
  - Communication
  - Electrical
  - Fire Protection
  - Mechanical
  - Gas
  - Plumbing

- X (Miscellaneous and hidden geometry)
  - CM—coordination
  - Hidden
  - Linked Models

Batch Printing and Exporting

When printing or exporting .dwf files or images, it is useful to use predefined batch sets which can be defined in the print and image export settings.

Materials

Materials in Revit are used for rendering, quantity takeoffs, and structural and thermal calculations. The materials in the VDC template include the physical and thermal characteristics and are named according to the Master Format System, with a six-digit number code. The benefit of this system is that organized materials

Figure 5.1.6
View/sheet set   12   6

always show up at the top of the materials list and any unvetted materials show up on the bottom.

There are five property categories in a Revit material:

1 Identity: Contains information regarding the material such as name, description and class. Product information is available but rarely used.
2 Graphics: This is the visual control center for color and patterns in line and shaded drawings in Revit.
3 Appearance: This is the visualization for renderings. The level of realism depends on the rendering system selected in Revit.
4 Physical Properties: This includes mechanical properties such as expansion coefficient, behavior, density, and strength relating to the specific material type.
5 Thermal Properties: There are nine thermal properties: behavior, thermal conductivity, specific heat, density, emissivity, permeability, porosity, reflectivity, and electrical resistivity.

Revit Shortcuts

Keyboard shortcuts are essential for speeding up repetitive tasks. Activating commands is faster with shortcuts. An experienced user employs shortcuts to keep from losing focus on a modeling task by searching for the button. Many commonly used Revit commands do not have a built-in shortcut. LiRo VDC created a custom shortcut file in order to keep the shortcuts consistent within the department.

Figure 5.1.7A–B
Materials windows in Revit

Starting a Revit Project

There is a set of procedures to complete in order to make sure that a project gets up and running correctly. The list below serves as a helpful guide. It is tailored specifically for the template used by the VDC team, but most of the steps could be applied to any project.

The first thing to note about the Revit template used by the VDC team is that it is not saved in the Revit template format (.rte) but rather in the Revit project file format (.rvt). For some reason, Revit template files cannot include worksets, which are essential for effective project organization. The LiRo VDC template depends highly on worksets for critical aspects of

Shortcuts

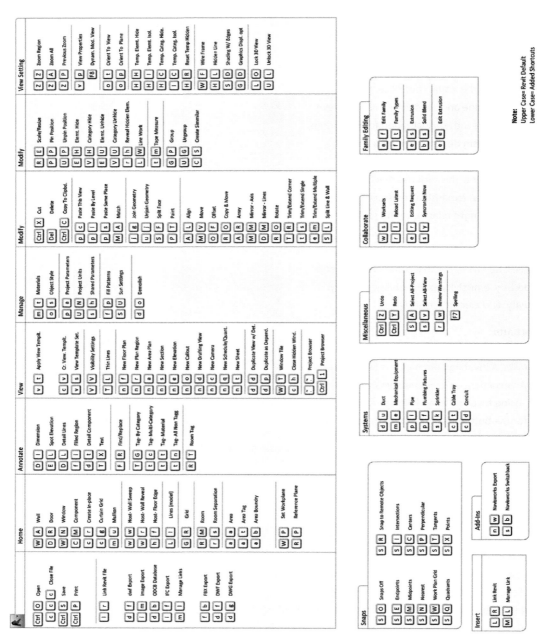

Figure 5.1.8
Keyboard shortcuts in Revit

project organization as many views and filters are dependent of pre-formatted worksets.

Essential Steps for Project Set-up

The necessary steps to set up a template are listed below:

1   In Revit, click open project and select the template file. Make sure to select "detach from central." Click save and select "maintain worksets."
2   Save the file with correct project name.
3   Define project name, number, and address.
4   Add client name, address, internal, and external team organization.
5   Enter energy settings (location, building type).
6   Document project team organization.
7   Hold team BIM meeting—goals and settings.
8   Plan Revit file linkage set-up.
9   Draw file linkage diagram.
10 Establish location coordinates and true and project north (project, survey).
11 Define title sheets.
12 Delete unneeded levels. Rename if needed.
13 Set building scope boxes.
14 Review and define possible phases.
15 Define consultant coordination set-up.

NAVISWORKS

Navisworks is used for coordination and clash detection. The software is not designed to incorporate as sophisticated a template workflow as Revit does; however, there are certain settings that are helpful to incorporate in Navisworks models. The main difference between the design and construction models is that during construction coordination, the project is further broken down, often by level or zones, while the design model usually houses the whole project. Below are some of the things that can be incorporated into each project.

Saved Viewpoints

*   Coordination
*   Exterior
*   Interior
*   Reviews
*   Visualization
*   Worksets

Selection Tree

The selection tree includes .dwf files that have been exported from Revit:

- ARC—Exterior.dwf
- ARC—Interior.dwf
- ARC—Finishes.dwf
- SRV—Communication.dwf
- SRV—Electrical.dwf
- SRV—Fire Protection.dwf
- SRV—Mechanical.dwf
- SRV—Gas.dwf
- SRV—Plumbing.dwf
- CM—Coordination.dwf

Search Sets

The search sets functionality in Navisworks is a powerful tool to search geometry based on different criteria. It is also used to define the correct geometry for clash detection. Most of the search sets used for coordination relate to helping with selection of different MEP pipe systems.

Clash Sets

Clash sets are used to run different clash scenarios following a hierarchy of importance. It provides a good guide in what sequence to run the tests.

Clash Hierarchy

1. Structure
2. Architecture
3. Mechanical Equipment Clearances
4. Mechanical Ducts
5. Sloping Pipes
6. Mechanical Pipes
7. Plumbing Pipes
8. Electrical Conduit
9. Fire Sprinkler
10. Fire Stand-pipe

LUMION

Lumion is a versatile program used to generate realistic walk-through animations of buildings. It can import .obj and Collada files from Sketchup, 3ds Max, and Revit. Plugins like Revit–Lumion Bridge help streamline the transfer of geometry from Revit to Lumion.

When working with particularly large projects, it is best to create dedicated 3-D views in Revit from which to export Collada files. These Revit views have visibility filters applied to them, often based on worksets. For large projects, it is recommended that worksets be dedicated to interiors, exteriors, finishes, structure, and also an additional "animation" workset (for geometry that does not exist in any design drawings, but intends only to bring the animation to life, like mock-up retail signage or display cases, for example). Changes made within one of these worksets can be exported from Revit and refreshed in the Lumion model more quickly than an export of the entire project model, because worksets are smaller and more manageable in size. Exporting a project in batches also allows the modeler to import them into the Lumion file as layers. The heavier the models and animation objects (people, trees, cars, etc.) visible in Lumion, the slower the frame rate of live rendering. But, sorting models and objects into layers that can be turned off when not in use keeps the pace of the program fast.

NEWFORMA PROJECT INFORMATION MANAGEMENT SYSTEM

A robust project information management system is essential to effectively manage project knowledge and material—everything from design documents, to e-mail chains, contractor contact details, and meeting minutes. In particular, it is best to use a system that has a web interface so that project data can be accessed on internet-enabled devices from the field. LiRo VDC uses Newforma primarily for project communication and generating coordination interference reports and notifications. Newforma logs and tracks data and helps automate information distribution and generate meeting minutes. The program has so much functionality for project management that it can be intimidating for new users.

SYNCHRO ORGANIZATION

Synchro does not use template files. Instead, the complexity of the project is managed through workflow and file set-up. A 4-D model consists of schedule data imported from Primavera or Microsoft Project, and 3-D geometrical data authored in VDC software such as Revit. Synchro is able not only to import the data into its simulation environment, but also to maintain a link to the original source of data so that updates to the geometry in the authoring software can be exported and synchronized to the linked schedule in Synchro without relinking. In the planning and execution phases of the project, schedules often change and the Primavera .xer file evolves with time. Project geometry may also be in flux.

The ability to synchronize with the original source ensures that the 4-D model reflects the most current available information. Every iteration of the schedule is saved inside the Synchro file as a separate version, and these versions can be compared by running simulations on multiple schedules simultaneously.

Synchro Set-up

There is a fair amount of set-up required to connect Primavera to Synchro, the details of which are available from the Synchro website. Once the connection is established between the software, the Primavera database can be accessed from within Synchro and imported schedules maintain their link to the Primavera file so that they can also be synchronized. It is important to maintain consistent activity IDs and schedule logic so that the established link inside Synchro between schedule activities and 3-D geometry remains unchanged.

Another way to set up the 4-D model is to import activities from Excel rather than directly linking to Primavera. This can be useful for expanding the workflow to include users who may not have access to Primavera. Exported Primavera schedules are saved in Excel and simply copied and pasted directly into the Synchro Gantt chart window. The Excel workflow is unidirectional, so synchronization with schedule changes is not possible. The Excel workflow is useful for simple, summarized 4-D studies; however, the full Primavera set-up is preferable, as it enables the team to use a single 4-D model throughout the course of construction.

Synchro Schedule Organization

An imported Primavera schedule contains all of the information embedded in Primavera, including its organizational layout, which can be used to organize the Synchro activity tree for ease of navigation. Applying work breakdown structure layouts arranges the schedule into a multilevel tree, with containers and sub-containers. It is also possible to organize the schedule according to user-defined filters in Synchro. Natively defined filters will remain in the Synchro file through the process of synchronization. The organization of the schedule is fundamental to the 4-D workflow and must be considered when the schedule is initially generated, in terms of which activities will be visualized in the 4-D model and to what level of detail.

Synchro Geometric Data and Resources

In Synchro software, objects contained in the model are referred to as resources. In a construction simulation, each of the construction activities requires a resource to be linked to it. A construction activity can contain a material, equipment, or human resource. When 3-D geometry is imported into Synchro, the user can specify the category of resource for the 3-D object. The process of generating a 4-D model consists of attaching different resources to construction schedule activities.

The Synchro workflow consists of organizing 3-D geometry in the information model into export sets according to the construction activities. These sets are imported into the Synchro model, assigned a single resource type, and directly linked to the corresponding construction activity. For example, all of the foundation piles could be exported in one bundle, and all of the pile caps exported in another, as the installation of piles is represented as a single construction activity and the installation of the pile caps as another in the project schedule. Since the project schedule drives the 4-D model, the configuration of 3-D exports should be organized according to an analysis of the schedule.

Use Profiles

One of the key components unique to the Synchro environment is the "use profile." Use profiles can define a state of action and the visual presentation of the action. Any activity can be assigned one of four states—install, remove, temporary, and maintain—which are controlled through the use profile. As the simulation progresses in time, if a 3-D element is mapped to an activity with an install-type use profile, the 3-D element will be highlighted during the duration of its corresponding activity and then display as permanently visible after the activity has completed, to simulate the action of installation. The remove use profile highlights the element during the activity, then renders it invisible after the activity has ended. The rendering options of the use profile include user-designed color and transparency. This ability to control both the color and transparency of the 3-D geometry with such specificity is a powerful tool for visualizing the project's construction. A project's use profiles should be set up after organization of the resource tree.

Distribution of 4-D Material

Synchro's animation capabilities include the production of quick, high-quality fly-throughs and schedule time lapses. The animation functionality is based on establishing viewpoints and keyframes on the animation controller timeline. Synchro automatically interpolates between two viewpoints for a seamless fly-through effect. The viewpoints are organized in a separate window, making it both functional and user friendly.

Animations can be output in a variety of video formats. The production workflow adopted by LiRo's VDC team consists of exporting animation in the form of still images and compiling them into a movie file in a post-processing environment, resulting in a higher quality video.

A free Synchro viewer enables distribution of a frozen, read-only 4-D model, simplifying sharing of 4-D information. These 4-D models must be carefully designed to convey the correct information, as the viewer can only view the model and cannot alter its set-up.

FILE MANAGEMENT: FOLDER STRUCTURE

A clear folder structure is extremely important for organizing the enormous amount of data created on a VDC project. The project folder structure template has grown into quite an extensive tree, which has been revised numerous times over the years in order to facilitate effective organization and workflow. For example, there are now many types of models and formats, and it is essential that some models are kept live and constantly changing, while others need to be kept simply as static records of a point in the process. Some folders are only shared with specific team members.

Several criteria form the very basic branches on the LiRo VDC folder structure. The first distinction is whether the information is shared with outside organizations or only used internally by the VDC team.

Project Folder Structure

When a new project is started, the template folder structure, which consists of an empty version of all the folders needed to store the project data, is copied into the new project location on the server. This template contains the many folders that are needed for all the different VDC services. While quite extensive, the folder structure is as simple as possible considering the complexities of running multiple models and collaborating with multiple team members.

The fundamental concept is a division between internal (dynamic) and external (static) files. Internal files are constantly being worked on while external files are either incoming or outgoing. Maintaining two separate folders for internal and external files is an effective way of handling user rights, as external folders can be shared with the broader project team.

- External (static files)
  - Published (outgoing)
    - Commissioning
    - Constructability
    - Cost Estimate
    - Management
    - Planning
    - Site Safety
    - Tracking
    - Visualization

- o Reference (incoming)
  - – 1 Survey
  - – 2 Photos
  - – 3 Design
  - – 4 Construction
  - – 5 Standards
- Internal (live files)
  - o Admin
    - – Contracts
    - – Project Set-up
    - – Proposals
    - – Time
  - o Authoring Models
    - – Revit
      - • Archive
      - • Files
      - • Links
    - – Other
  - o Library
    - – Architectural
    - – Engineering
    - – Visualization
  - o VDC Services
    - – Commissioning
    - – Constructability
      - • Coordination
        - o Design
          - – Models
          - – Links
        - o Construction
          - – Models
          - – Links
        - o Clash Detection
      - • vRFI
    - – Cost Estimate
    - – Management
    - – Planning
    - – Prototyping
    - – Site Safety
    - – Tracking
    - – Visualization

5.2 VDC Specifications

VDC is still a new concept to quite a few people who are not specialists in the field. To ensure everybody on a project has the same expectations, goals, and a clear understanding of how to achieve them, it is essential to have solid, clear, and easy-to-understand VDC specifications.

Depending on the extent to which VDC should be used on a project, specifications can vary in complexity and depth. Some are extensive, spanning from pre-design to final operations, while others focus on specific phases or trades. Certain aspects are common across every specification. They should all include a scope outline, definitions, the required submittals, software and hardware standards, modeling requirements, reference standards, execution, and trades requirements.

The VDC specification is a document designed to describe the required end product, as well as the framework by which the final product will be developed. The VDC specification also describes the types of VDC products the contractor is required to deliver. It is a supplement to the signed contract between two entities, and might vary depending on their relationship. LiRo's VDC group has experience in working with the team to produce VDC specifications from the perspective of the owner and construction manager on a series of projects.

As VDC is a disruptive practice in the industry, naturally some in the construction community are trepidatious regarding its implementation. For example, decision makers on the ESA project expressed concerns about the integration of VDC processes into their conventional construction process, because of their unfamiliarity with VDC. Opinions on the project ranged from strong advocates of VDC to strong opponents. In this case, one of management's main concerns was whether the contractor's ability to develop VDC processes was mature enough to aid in construction and not slow it down. The VDC specifications needed to be developed to describe and clearly outline all of the requirements for the contractor.

There are various examples of VDC specifications available from different sources such as the American Institute of Architects and the US Army Corp of Engineers. These documents are very helpful, but they usually contain general information about VDC and try to cover a variety of scenarios. It is essential to thoroughly analyze these documents and only adopt the portions applicable to the project specific. It is important to include all of the relevant information from outside sources into the specification itself, rather than reference supplemental documents.

However, some established standards may be referenced, such as the Uniformat or Master format.

All VDC products required by the client need to be thoroughly described in specification documents. Though the information model is at the center of the VDC process, the construction process comprises different stages and involves a whole ecosystem of different products. The specification needs to outline all of these products and describe in great detail the types of VDC products that clients expect contractors to develop and implement. It is also essential to include detailed descriptions of the professionals expected to develop and manage the VDC ecosystem and clearly define the role of the client.

To ensure the success of VDC implementation, it is necessary to describe the relationship between the VDC representatives on the client side and the VDC managers on the contractor side. As such, it is of vital importance that the contractor create a VDC implementation plan shortly after the award of the contract. Since the client cannot dictate means and methods to the contractor, the client should require a full description of the VDC applications in the VDC Implementation Plan. After the contractor submits an implementation plan, the client will have a chance to evaluate and make comments during the implementation process.

If the goal of a project is to develop a specific type of product—a facility management model, for example—the specification needs to outline the data format and level of development (LOD) expected at delivery. The document needs to be rigid and detailed to ensure successful development and implementation while also taking into account the evolving nature of the VDC industry. Below is an example of a typical VDC specification.

EXAMPLE OF BUILDING INFORMATION MODELING (VDC) SPECIFICATION

PART 1 General

1.01 SECTION INCLUDES

A. Requirements for furnishing all labor, materials, tools, and equipment, and performing all operations necessary for Building Information Modeling and the preparation, update, submittal, and management of Building Information Models.

B. Definitions
    1   Building Information Modeling: Process by which data-rich digital representations of the physical and functional characteristics of a facility are created and maintained.

2   Building Information Model (BIM): A three-dimensional and data-enabled digital representation of the physical and functional characteristics of a facility.

3   Virtual Design and Construction (VDC): The development and implementation of technology and BIM-based processes to achieve benefits to project production and outcomes.

4   Existing Condition Model(s): A BIM of existing features of a facility created from a point cloud(s) generated by a laser scan, or from record documents or as-built documents.

5   CM Architecture Design Intent Model: A BIM developed by the Construction Manager from the architectural design drawings, not intended for construction, but provided to the Contractor for reference.

6   Contractor Discipline Model(s): A BIM or series of BIMs developed by the Contractor for each required discipline.

7   Contractor Master Discipline Model: A BIM comprised of linked Contractor Discipline Models for a single discipline.

8   Contractor Master Model: A BIM comprised of all linked Contractor Discipline Master Models.

9   As-Built Model(s): Contractor Discipline Models updated to reflect actual shape, size, and position of constructed elements.

10 Master As-Built Model: A BIM comprised of all linked As-Built Models.

11 Facility Data or Attribute Data: Any of the data which is represented by, used for, or associated with the BIMs for the entire lifecycle of the facility, including design, construction, and facility management by the Client.

12 BIM Execution Plan (BEP): A document, which communicates all information, related to the BIM process. The BEP includes contact information of BIM leads, BIM goals and uses, BIM software, BIM Standards, Model Management protocols, and level of Development (LOD).

13 BIM Uses: The output, utilizations, and processes facilitated by or extracted from the BIM.

1.02 REFERENCED SECTIONS [separate sections not included in this book]

A.   Section 01330—Submittal Procedures

B.   Section 01720—Surveying

C.   Section 01750—Asset Management

1.03 CITED STANDARDS

A. Level of development (LOD) Specification 2013
https://bimforum.org/lodhttps://bimforum.org/lod/

B. National BIM Standard V2 www.nationalbimstandard.org

C. Construction Operations Building Exchange (COBie)
www.wbdg.org/resources/cobie.php

1.04 NOTED RESTRICTIONS

A. BIM files: BIM files developed and submitted by the Contractor shall be
in Autodesk 2015 Revit and Navisworks 2015 format as specified
herein unless otherwise directed by the Construction Manager.

1.05 SUBMITTALS

A. BIM Manager Resumé: Within 30 days of the date of Notice to Proceed,
the Contractor shall provide the resumé of the BIM Manager to the
Construction Manager for review.

B. Electronic Data Disclaimer: Within 30 days of the date of Notice to
Proceed, the Contractor shall provide a completed Electronic Data
Disclaimer to the Construction Manager.

C. BIM Software and Hardware List: Within 30 days of the date of Notice
to Proceed, the Contractor shall provide a list of proposed BIM-related
software and hardware, along with any requests for waiver, to the
Construction Manager for review.

D. BIM Execution Plan (BEP): Within 30 days of the date of Notice to
Proceed, the Contractor shall provide the initial release of the BEP to
the Construction Manager for review. The Contractor shall provide
necessary periodic updates to BEP to the Construction Manager for
review.

1.06 DELIVERABLES

A. Contractor BIMs: Beginning 60 days after Notice to Proceed and at
intervals of one week thereafter until Substantial Completion, the
Contractor shall submit current versions of Contractor Discipline,
Master Discipline, and Master Models to the Construction Manager for
review.

B. 3-D Coordination Meeting Minutes and Clash Status Matrix: Beginning
60 days after Notice to Proceed and at intervals of one week thereafter
until Substantial Completion, the Contractor shall submit within two
business days of meeting date minutes of 3-D Coordination Meetings

along with updated Clash Status Matrix to the Construction Manager for review.

C. Contractor As-Built BIMs: Within 60 days of completed field installation, the Contractor shall provide updates to As-Built models and Master As-Built model, containing recently constructed elements, to the Construction Manager for review.

D. COBie Spreadsheet: The Contractor shall provide a completed COBie spreadsheet for equipment and assemblies as required by the Client Asset Management specification.

PART 2 Products

A. The Contractor shall provide a list of proposed BIM software to the Client prior to start of modeling. BIM software currently acceptable to the Client is as specified herein. Deviations from this software require explicit waiver from the Client.

B. Unless otherwise agreed to, the Contractor shall use Autodesk Revit Architecture 2015 BIM Authoring Software to develop and update Contractor Discipline and Master Discipline Models for Architectural discipline.

C. Unless otherwise agreed to, the Contractor shall use Autodesk Revit Structure 2015 BIM Authoring Software to develop and update Contractor Discipline and Master Discipline Models for Structural discipline.

D. Unless otherwise agreed to, the Contractor shall use Autodesk Revit MEP 2015 BIM Authoring Software to develop and update Contractor Discipline and Master Discipline Models for Mechanical, Electrical, Plumbing, IT/COMM, and Fire Protection disciplines.

E. Unless otherwise agreed to, the Contractor shall use Autodesk Revit Architecture 2015 BIM Authoring Software to develop and update As-Built Architectural Discipline Models.

F. Unless otherwise agreed to, the Contractor shall use Autodesk Revit Structure 2015 BIM Authoring Software to develop and update As-Built Structural Discipline Models.

G. Unless otherwise agreed to, the Contractor shall use Autodesk Revit MEP 2015 BIM Authoring Software to develop and update As-Built Mechanical, Electrical, Plumbing, IT/COMM, and Fire Protection Discipline Models.

H. Unless otherwise agreed to, the Contractor shall use Autodesk Navisworks Manage 2015 or Autodesk BIM 360 Glue to develop and update federated Master Model and perform 3-D Coordination/Clash Detection.

PART 3 Execution

3.01 GENERAL

A. The requirements and provisions of Section 01330 "SUBMITTAL
    PROCEDURES" shall apply in addition to the requirements of this
    section.

B. The Contractor shall create and maintain comprehensive,
    three-dimensional, parametric models.

C. The Contractor shall develop and continuously maintain coordinated,
    clash-free Contractor Discipline Models of architectural, civil,
    structural, electrical, mechanical, fire protection, IT, communications,
    security, and plumbing disciplines as specified herein, prior to
    production of shop drawings, procurement, installation, and on-site
    construction.

D. The Contractor shall report all conflicts, omissions, and discrepancies
    found in the Contract Documents while developing Contractor
    Discipline Models immediately to the Construction Manager for review
    and resolution.

E. Within two weeks of commencing Trade discipline development, the
    Contractor shall hold weekly 3-D coordination/Clash analysis meetings
    to identify and resolve clashes within and between Contractor
    Discipline Models.

F. The Contractor shall update all Contractor Discipline Models to
    reflect current Trade discipline development at least 24 hours prior to
    scheduled weekly 3-D coordination/Clash analysis meetings.

G. The Contractor shall aggregate, weekly, all current Contractor
    Discipline Models to create a Contractor Master Model prior to
    scheduled weekly 3-D coordination/Clash analysis meetings.

H. The Contractor shall perform detailed Clash analysis of the Contractor
    Master Model during each weekly 3-D coordination/Clash analysis
    meeting. Clash tests defined in the BEP shall be run and all resulting
    clashes reviewed, reconciled, and tracked.

I. The Contractor shall develop and maintain detailed Clash reports of
    each weekly Clash analysis and shall provide same along with the
    Contractor Master Model in rvt and nwd format to the Construction
    Manager.

J. The Contractor shall generate and update Shop Drawings as specified
    elsewhere in the Contract Documents, from the coordinated, clash-
    free Contractor Discipline Models as specified herein.

K. The Contractor shall verify that the coordinated, clash-free Contractor Discipline Models possess required spatial and dimensional accuracy, comprehensive and detailed content to drive fabrication processes, and generate points for automated field layout stations.

L. The Contractor shall install all systems represented in the coordinated, clash-free Contractor Discipline Models as modeled and as specified elsewhere within the Contract Documents.

M. The Contractor shall also develop a dimensionally accurate As-Built Model of structure, architecture, and MEP disciplines that embed Agency required Asset Management data compatible with the COBie format.

3.02 BIM MANAGER

A. The Contractor shall designate a qualified, experienced BIM Manager, subject to review by the Construction Manager, to manage and lead the Building Information Modeling process and act as primary interface between all Trade Subcontractors, Construction Manager, and Client. Duties include enforcement of applicable Client Standards for BIM and model management tasks necessary to ensure the quality of BIMs throughout the construction phase.

B. Minimum BIM Manager qualifications shall be as follows:

1   Minimum 10 years' experience on construction projects of similar size, scope, and complexity.

2   Minimum three years' experience in the management of VDC processes and the development and use of 3-D Models for construction. Experience must include at least one project of similar size and complexity.

3   In-depth knowledge of current VDC methodologies and BIM software.

4   In-depth knowledge of BIM standards specified herein.

5   Good written and verbal communication and training skills.

6   Strong teaching and coaching skills.

7   Flexibility and ability to understand and implement BIM standards with a multifaceted construction team and manage the delivery of quality products throughout the process of construction.

3.03 BIM EXECUTION PLAN

A. The Contractor shall prepare the BIM Execution Plan (BEP), using a BEP format approved by the Construction Manager, for the entire construction process, including the development and maintenance of As-Built Models and population of asset management attribute data properties. The BEP shall be submitted to the Construction Manager for approval prior to commencement of any modeling and shall include, at a minimum, the following items:

1   General Contractor Model requirements.
2   Model partitioning, organization, and structure.
3   Model units and local and absolute coordinate systems and vertical data.
4   Software Versioning, File Format, and Naming for all Models.
5   List of BIMs to be developed.
6   Organization and Contact Information of BIM coordinators and authors.
7   Any Required Software Object Enablers.
8   Process for generating COBie spreadsheets of asset management data from As-Built Construction Model.
9   List of Asset Management data properties included in the Construction Model.
10 Process for maintaining As-Built Models on an ongoing basis, including incorporation of layout data.
11 Description of data storage and data exchange, sharing, viewing, drafting protocols, and updating of models and supporting documentation.
12 Plan for BIM quality control, including schedule for quality control checking and monthly reporting on the integrity of the Models to the Construction Manager.
13 Process for generating shop drawings from coordinated clash-free Contractor Discipline Models.
14 Process for tracking changes to Contractor Models.
15 The Structure and Organization of the Models shall be approved by the Construction Manager.
16 Coordination/Clash Detection Process.
17 Coordination/Clash Detection Schedule.
18 Processes for using BIM-based layout tools to lay out and verify positioning of installation.
19 List of Clashes to be run during 3-D Coordination session, including tolerances for each component.

B. The Contractor shall submit, within 30 days of Notice to Proceed, the BEP to the Construction Manager for review. The Construction Manager shall confirm acceptability of the BEP or advice as to additional processes and/or activities necessary to be incorporated. If modifications are required, the Contractor shall execute the modifications and resubmit the final BEP to the Construction Manager.

3.04 BIM MEETINGS

A. The Contractor shall attend a BIM Kickoff Meeting, scheduled and conducted by the Construction Manager, within 30 days of Notice to Proceed. The meeting shall review BIM expectations and goals, requirements, schedule, and deliverables for the Project.

B. The Contractor shall attend a BEP Review Meeting, scheduled and conducted by the Construction Manager, within 30 days of Notice to Proceed. The Contractor, Construction Manager, and Client key personnel shall attend this meeting and review the proposed BEP in detail.

C. The Contractor shall attend monthly BIM Program meetings with the Construction Manager to review overall progress and any issues identified during the past month.

D. The Construction Manager shall prepare and distribute minutes of meetings to all participants and concerned parties.

3.05 CONTRACTOR MODEL REQUIREMENTS

A. The Contractor shall develop all Contractor Discipline Models for architecture, civil, structural, electrical, mechanical, plumbing, fire protection, IT/communication, and security disciplines to include all required building systems and components to 2013 BIM Forum LOD 400.

B. The Contractor shall define the Global coordinate system and units for all Contractor Models to use survey coordinate system datum as specified Section 01720 "SURVEYING" and defined in the Existing Condition Model.

C. The Contractor shall define the local coordinate system and units for all Contractor Models as specified in the BEP.

D. The Contractor shall reference grids and levels contained in the Existing Conditions Model. Creation of additional proposed grids and levels shall be coordinated with the Construction Manager.

E. The Contractor shall define subsystems by system type or clearly named layers based on a model color standard coordinated with the Construction Manager.

F. The Contractor shall model clearance areas required for equipment access, public safety, code, etc. as three-dimensional shapes assigned a distinct, easily identifiable name.

G. The Contractor shall purge models of unused elements and external links prior to distribution and submittal.

H. The Contractor shall assign components created in Contractor Models the correct CSI 2010 Uniformat II category.

I. The Contractor shall develop component-type names to include the function of the component (e.g. Interior versus exterior).

J. The Contractor shall assign correct materials within composite components such as walls, floors, ceilings, and roofs, to facilitate classification of 3-D objects into correct cost groups. For example, a floor component with a metal deck shall assign the correct metal material in the structure definition of the component.

K. The Contractor shall name all components using a specific and clear naming convention. The component naming convention shall be as defined in the BEP.

L. The Contractor shall assign the correct level to all columns, walls, and other component objects.

M. The Contractor shall model columns, walls, and similar components spanning multiple levels as separate components on each level.

N. The Contractor shall define room finishes as parameters for base finish, ceiling finish, wall finish, and floor finish.

O. The Contractor shall define all enclosed spaces in Discipline Architectural Models as rooms.

P. The Contractor, at a minimum, shall develop Contractor Discipline Models containing the model elements as specified in Appendix A, "Contractor Model Contents."

3.06 ASSET MANAGEMENT DATA REQUIREMENTS

A. The requirements and provisions of Section 01760 "ASSET MANAGEMENT" shall apply in addition to the requirements of this section.

B. The Contractor shall populate a COBie v.2.26 spreadsheet with data extracted from Contractor Discipline Models for electrical, mechanical, plumbing, fire protection, and vertical circulation disciplines. Extracted data shall be as defined in the following table.

Asset Data Property	Spreadsheet Tab	Column
Asset Tag	Component	Tag Number
Asset Description	Type	Name
Asset Type	Type	Asset Type
Asset Group	Type	Category
Facility Identification	Facility	Site Description
Building Name	Facility	Name
Floor Identification	Floor	Name
Location Identification	Component	Space
Manufacturer	Type	Manufacturer
Model Type	Model	Number
Serial Number	Component	Serial Number
Warranty Start Date	Component	Warranty Start Date
Warranty Expiry Date	Type	Warranty Duration Parts
Linkage Data Field	TBD	TBD

CONTRACTOR DISCIPLINE MODEL CONTENTS

1. HVAC

DUCTWORK

- Main duct runs
- Secondary duct runs (from main duct runs to fixtures and in-line equipment)
- Fixtures (diffusers, grilles, registers, fume hoods)
- Connections to fixtures
- Insulation
- Required equipment clearance/access zones modeled as separate components

EQUIPMENT AND ACCESSORIES

- Air terminal boxes
- Sound attenuators
- Exhaust fans
- In-duct heating and cooling coils
- Humidifiers
- Fire and smoke dampers

- Access panels
- Hangers/support
- Equipment pads

2. PLUMBING

PIPING

- Piping (hot water, chilled water, condensate, etc.)
- Connections to equipment and fixtures
- Insulation

EQUIPMENT AND ACCESSORIES

- Pumps
- Cabinet unit heaters
- Hot water unit heaters
- Radiators
- Fin tubes
- Radiant ceiling/wall panels and floors
- Un-ducted fan coil units
- Valves
- Air handling units
- Sound traps
- Main fans and blowers
- Tanks
- Main sound attenuators
- Main water treatment units
- Main pumps
- Split system air conditioners
- BAS devices and fittings
- Variable frequency drives
- Heat recovery units
- Filter housings
- Compressors
- Dryers
- Sinks
- Service sinks
- Fixture support carriers
- Showers
- Floor sinks
- Floor drains
- Roof drains

3. FIRE PROTECTION

- Piping
- Sprinkler mains
- Sprinkler branch piping
- Sprinkler pipe insulation
- Sprinkler heads
- Stand-pipes and risers
- Instrumentation and controls
- Required equipment clearance/access zones modeled as separate components
- Fire pumps
- Tanks
- Valves
- Seismic bracing

4. ELECTRICAL

- Power distribution
- Cable tray
- Bus duct
- Pull boxes
- Required equipment clearance/access zones modeled as separate components
- Equipment and accessories
- Transformers
- Panel boards
- Equipment pads
- Power feeds
- Switch gear
- Specialty systems (generators, UPS, etc.)
- Fixtures
- Lighting fixtures—lay-in and ceiling mounted
- Lighting fixtures—wall mounted and vertical
- Lighting fixtures—exterior
- Emergency exit signs

5. STRUCTURAL

- Structural columns, beams, joists, trusses, and main braces
- Secondary elements (braces, connection members and plates, etc.)
- Base plates
- Gusset plates
- Lintels
- Kickers

- Clip angels
- Slabs and openings

6. ARCHITECTURAL

- Finish floors
- Interior and exterior walls
- Penetrations
- Ceilings
- Column covers
- Curtain walls
- Doors
- Plumbing fixtures
- Rooms
- Lighting fixtures
- Railings
- Ramps
- Roofs
- Specialty equipment
- Stairs
- Windows
- Levels
- Grids

7. VERTICAL CIRCULATION

- Escalators
- Elevators

8. CIVIL

- Site features

5.3 VDC Implementation Plan

Implementing change can be a challenging task, especially when dealing with large organizations and professionals with years of experience based on certain engrained procedures. In order for VDC methodology to be effective, guidance must be provided in the form of a logically structured document that outlines a clear plan.

The VDC implementation plan is a direct response to the VDC specifications and serves as a road map for the successful execution of the methodology. The VDC implementation plan must be submitted by the party responsible for managing VDC. The responsible party will likely be the design team during the design phase and the contractor during the construction phase. Teams may encounter situations not outlined in the plan as a project progresses; therefore an implementation plan needs to provide a strong foundation that is also adaptable to a project's unique circumstances.

The CM needs to guide the project team through the process of implementation, making sure that all decisions are made with a holistic view of all participants in mind.

For the City Point Project, the role of the LiRo VDC team included managing and coordinating all the design and construction models. Therefore the LiRo team was responsible for crafting the VDC implementation plan. The main purpose of the plan was to guide the project through the process of systems coordination.

Because construction is a dynamic process, not all circumstances can be envisioned at the beginning of the VDC effort; therefore it is necessary to update VDC implementation plans periodically. However, at the commencement of the VDC effort, a strong foundation needs to be laid out and agreed upon. The City Point construction phase implementation plan is offered below as an example to provide a foundation for VDC implementation plans that can be tailored to the specific needs of different projects.

GENERIC VDC IMPLEMENTATION PLAN

General

Definitions

- vRFI—Virtual Requests for Information: A markup that derives out of the process of modeling highlighting issues to be resolved, or areas that require more information for successful construction.

- Coordination Model: Central model that holds the structural, architectural, and systems sub-models. Managed by the LiRo virtual construction coordinator and used to generate vRFIs and clash detection reports.
- Design Intent Model: LOD 300 level model which includes architecture and structure as depicted in the construction documents.
- Construction Model(s): Models for each system to include all elements, equipment, fittings, etc., accurate in size, shape, location, quantity, and orientation with complete fabrication assembly and detailing info as required in LOD 400 (AIA-document E202). To be modeled by system by level.
- Distribution Model: Current Navisworks version of the coordination model saved in Newforma shared folder. Contains all of the vRFI and minor clash views.
- Systems Coordination Drawing Set: The system coordination drawing set consists of floor plans generated by level and by system from the coordination model. Plans contain the 100 percent CD CAD set with the design intent model and the construction models overlaid.
- Shared Folders:
  a. Construction models shall be uploaded by subcontractors into folders named for each subcontractor on the project.
  b. "Design Intent" folder contains a "DWG" folder with the design intent model saved in 3-D DWG format by system by level, and a "Revit" folder which contains the complete design intent model saved in Revit format.
  c. "Podium Coordination" folder contains the most recent coordination model saved in Navisworks format.

Requirements

- All participants shall submit construction models in a file format compatible with Navisworks 2013.
- All participants are required to install at a minimum the free Navisworks viewer to review coordination model.
- Construction model(s) shall be delivered free of links, unnecessary views, abandoned designs, etc.
- Construction models shall be modeled using the prescribed project origin.
- Participants shall update construction models in a timely fashion, as discussed and agreed upon in coordination meetings.

- Model updates shall be submitted to the LiRo virtual construction coordinator in a timely fashion prior to all scheduled coordination meetings, to enable clash detection and review.
- Models will be distributed through a shared folder on the Newforma Info Exchange website.

Software

- Autodesk Navisworks Freedom: Free viewer for Navisworks models available for download from the Autodesk website.
- Newforma Project Center: Accessed via web browser. Participants will receive an invitation email.

MODEL REQUIREMENTS

Common Coordinate System and Units

Project Origin
All drawings and models are to be submitted using the project coordinates, elevations, and units (feet and inches) which are established in the Revit design intent model. No shifting offset or rotations will be permitted.

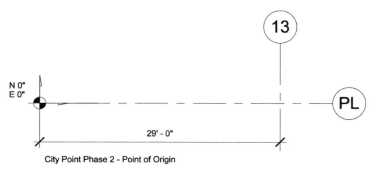

City Point Phase 2 - Point of Origin

Figure 5.3.1
Point of origin in City Point Phase 2

Elevation Mark
All models submitted shall include at least two reference grid lines and an elevation mark shown on the project base point provided to participants to verify that the construction models match the design intent model coordinates.

Figure 5.3.2
Elevation mark

Model Partitioning

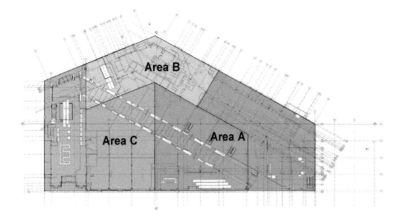

Figure 5.3.3
An example of model partitioning

Area A is bounded by grid lines: A | 0.8 | 8 | R.1 | 5.1 | U.1
Area B is bounded by grid lines: R.1 | 5.1 | U | P | 11 | L
Area C is bounded by grid lines: P | 13 | A | 8 | L | 11

Construction Model Requirements

- Construction models shall use the project origin and levels as established by LiRo in the design intent, coordination, and distribution models.
- Subsystems shall be saved on distinct, clearly named layers using the coordination model color standards.
- Clearance areas required for equipment access shall be modeled as boxes and saved on a distinct, clearly named layer.

FILE-NAMING CONVENTION

Construction model files submitted to the LiRo virtual construction coordinator shall be submitted as separate files by trade and by level and named as follows:

CityPoint_TRADE_LEVEL_Date

For example, the plumbing model for level 5 submitted July 1, 2013 would be named as CityPoint_PLU_LV5_20130701.

The following three-letter abbreviations must be used to designate trade and level:

Level
- LSC   Sub-Concourse
- LCN   Concourse

Color Scheme - By Trade		
Trade / Tower	Color Code	RGB Value
MEC - Mechanical	Green (Dark)	000-155-000
PLU - Plumbing	Purple	175-046-190
FPT - Fire Protection	Red (Dark)	193-000-000
ELC - Electrical	Yellow (Light)	255-255-000
COM - Communication	Orange	240-100-030
Tower 1	Blue (Dark)	030-060-160
Tower 2	Yellow (Dark)	180-180-000

Color Scheme - By System			
Trade	System	Color Code	RGB Value
MEC	Return Air	Red (Light)	255-062-062
	Supply Air	Blue (Light)	000-255-255
	Exhaust Air	Orange	240-100-030
	Makeup Air	Green (Light)	050-255-050
	Outside Air	Blue (Medium)	079-112-223
	Chilled Water Return	Blue (Light)	000-255-255
	Chilled Water Supply	Blue (Light)	000-255-255
	Condensate Drain	Blue (Medium)	079-112-223
	Condenser Water Return	Green (Light)	050-255-050
	Condenser Water Supply	Green (Light)	050-255-050
	Heating Hot Water Return	Red (Light)	255-062-062
	Heating Hot Water Supply	Red (Light)	255-062-062
	Natural Gas	Yellow (Light)	255-255-000
	Other	Black	060-060-060
	Mechanical Equipment	Green (Dark)	000-155-000
PLU	Domestic Cold Water	Blue (Light)	000-255-255
	Domestic Hot Water	Red (Light)	255-062-062
	Domestic Hot Water Return	Red (Dark)	193-000-000
	Ejector Discharge	Purple	175-046-190
	Other	Black	060-060-060
	Sanitary (Blackwater)	Brown	139-069-019
	Storm Water	Green (Dark)	000-155-000
	Vent	Green (Light)	050-255-050
	Waste	Orange	240-100-030
	Plumbing Fixtures / Eqt	Purple	175-046-190
FPT	Fire Standpipe	Red (Light)	255-062-062
	Sprinkler Line	Blue (Medium)	079-112-223
	Sprinklers/Fire Protection Eqt	Red (Dark)	193-000-000
ELC	General Conduits	Yellow (Light)	255-255-000
	Electrical Equipment	Yellow (Light)	255-255-000
COM	General Conduits	Orange	240-100-030
	Communication Equipment	Orange	240-100-030

Figure 5.3.4
Coordination model color standards

- LV1   1st Floor (Ground)
- MEZ   Mezzanine
- LV2   2nd Floor
- LV3   3rd Floor
- LV4   4th Floor
- LV5   5th Floor

Trade

- MDC   HVAC Duct
- MPI   HVAC Piping
- PLU   Plumbing
- SDP   Fire Stand-pipe
- SPK   Fire Sprinkler
- ELC   Electric
- FAM   Fire Alarm
- BMS   BMS
- COM   Communication
- SEC   Security
- STR   Structure
- ARC   Architecture

vRFI Naming Convention

Specific issues and areas of conflict will be saved as views in the Navisworks coordination model and distributed to the participants as vRFIs (Virtual Requests for Information). vRFIs shall be organized by level and area and named with the following convention:

LEVEL_AREA_vRFI#_GRIDLINES_TRADE(S)

For example, the second vRFI on level 2 for a conflict between structure and plumbing at grid lines B and 3.5 in area A would be named as LV2_ARA_002_B0.0-03.5_STR_PLU

LOD-400 (AIA E202) CONSTRUCTION MODEL REQUIREMENTS
For the list of LOD-400 model requirements, see Chapter 5, Section 5.5.

COORDINATION WORKFLOW

Design Intent Coordination Workflow

1 VDC analyzes design intent model for areas of conflict and creates reference Navisworks views.
2 Conflicts are categorized into critical and minor categories.
3 Critical issues are encapsulated in vRFI notifications and logged and submitted to the coordination team via Newforma Project Center.
4 vRFIs resolutions are discussed in weekly team coordination meetings. As vRFIs are resolved, they are designated closed.
5 LiRo is responsible for updating the design intent systems model to reflect vRFI resolutions.
6 Minor issues are documented in the Navisworks coordination model by LiRo and are not remodeled or resolved by LiRo. Subcontractors are responsible for taking minor issues into account in the development of construction models.

Subcontractor Coordination Workflow

1 Subcontractors receive Navisworks coordination model with vRFI views for reference in developing construction models.
2 Subcontractors receive design intent model saved by level in DWG and Revit format.
3 Subcontractors are required to develop construction models using the defined project origin and taking into account design intent vRFI resolutions and design intent minor conflicts, as documented in the Navisworks coordination model.
4 Subcontractor construction models are uploaded to LiRo and imported into the coordination model for conflict detection, as per the clash resolution hierarchy.
5 Coordination shall be managed by level and by system in accordance with the clash process diagram. Clash detection reports will be run on the models as per the clash detection matrix, and clashes reviewed for critical and minor conflicts.
6 Minor and major conflicts are saved as views in the Navisworks distribution model and referred to subcontractors.
7 Minor clashes will not be individually tracked; each subcontractor is responsible for taking into account such clashes and resolving them in the revised construction model.
8 Major conflicts are discussed and resolved in subcontractor coordination meetings, schedule to be determined.

248   REFERENCE DOCUMENTS

CITY POINT SUBCONTRACTOR COORDINATION CLASH MATRIX

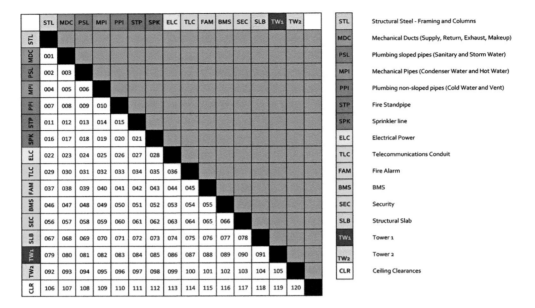

	STL	MDC	PSL	MPI	PPI	STP	SPK	ELC	TLC	FAM	BMS	SEC	SLB	TW1	TW2	
STL	■															
MDC	001	■														
PSL	002	003	■													
MPI	004	005	006	■												
PPI	007	008	009	010	■											
STP	011	012	013	014	015	■										
SPK	016	017	018	019	020	021	■									
ELC	022	023	024	025	026	027	028	■								
TLC	029	030	031	032	033	034	035	036	■							
FAM	037	038	039	040	041	042	043	044	045	■						
BMS	046	047	048	049	050	051	052	053	054	055	■					
SEC	056	057	058	059	060	061	062	063	064	065	066	■				
SLB	067	068	069	070	071	072	073	074	075	076	077	078	■			
TW1	079	080	081	082	083	084	085	086	087	088	089	090	091	■		
TW2	092	093	094	095	096	097	098	099	100	101	102	103	104	105	■	
CLR	106	107	108	109	110	111	112	113	114	115	116	117	118	119	120	■

STL	Structural Steel - Framing and Columns
MDC	Mechanical Ducts (Supply, Return, Exhaust, Makeup)
PSL	Plumbing sloped pipes (Sanitary and Storm Water)
MPI	Mechanical Pipes (Condenser Water and Hot Water)
PPI	Plumbing non-sloped pipes (Cold Water and Vent)
STP	Fire Standpipe
SPK	Sprinkler line
ELC	Electrical Power
TLC	Telecommunications Conduit
FAM	Fire Alarm
BMS	BMS
SEC	Security
SLB	Structural Slab
TW1	Tower 1
TW2	Tower 2
CLR	Ceiling Clearances

Figure 5.3.5
A clash matrix

Clash Resolution Hierarchy

1 Structure
2 Architecture
3 Mechanical Equipment Clearances
4 Mechanical Ducts
5 Sloped Pipes
6 Mechanical Pipes
7 Plumbing Pipes
8 Electrical and Communication Conduit
9 Fire Sprinkler
10 Fire Stand-pipes

Clash Tolerances by System

MDC	HVAC Duct	2"	MPI	HVAC Piping	2"
PLU	Plumbing	2"	SDP	Fire Stand-pipe	2"
SPK	Fire Sprinkler	2"	ELC	Electric	2"
FAM	Fire Alarm	2"	BMS	BMS	2"
COM	Communication	2"	SEC	Security	2"
STR	Structure	2"	ARC	Architecture	2"
MEC	Mechanical Equipment Clearance	2"	SPI	Sloped Pipes	2"

5.4 Classification Systems

Building coding standards are essential in order to keep a project organized and effectively exchange data between different models. At LiRo VDC, data standards are at the core of how work is performed. Currently, three different coding systems are used on VDC projects. The Unformat and Master formats were developed prior to the existence of information modeling, while the OmniClass was designed specifically to manage BIM data.

The UniFormat system is an effective way to organize a construction project, especially one utilizing VDC. In the Revit environment, it is referred to as the assembly code system. Design teams often do not assign assembly codes to their components, but this is essential in order to use the design intent model to run cost estimation and clash detection and help with 4-D scheduling.

The UniFormat system was developed by the American Institute of Architects (AIA) together with the U.S. General Service Administration in the 1970s.[1] It is used mainly in the United States and Canada and loosely follows the sequence of construction and how groups of components are put together, hence the term assembly system. From a modeling perspective, UniFormat is more straightforward than MasterFormat (CSI), which focuses on material specifications. OmniClass is a more recent, all-encompassing format that includes processes, standards, methods, and objects. OmniClass is embraced by the National BIM standard, as it codes almost every component and process. However, it results in extensive lists of two-digit numbers, indicating that it is not really designed for human workflows, but rather as a coding system for databases. Until OmniClass incorporates more automation into how those numbers are applied, it is too cumbersome and time consuming to work with. Revit has a far more streamlined process for using UniFormat, which is why LiRo VDC chose it as a standard for coding building components. UniFormat is also used by the LOD 2013 standard. In time, however, there will probably be a transition to OmniClass.

UNIFORMAT

The most recent version of UniFormat was published in 2010 and is a trademark of CSI and CSC. It was primarily developed as a standard for cost estimating to be used during the construction phase. It focuses on the objects of construction more than on processes, as other standards do.

The system is broken down into three levels. Level one is alphabetically broken down into major group elements. The second level is group elements, which is further broken down into a third level consisting of individual elements. There is sometimes a fourth or even a fifth level for some specific components, but knowing the three levels provides sufficient breakdown to categorize the building components.

For more information regarding UniFormat, see: www.wbdg.org/ndbm/uniformat.php

When modeling in Revit, all components are pre-wired to the most common UniFormat categories relating to the specific Revit categories. For example, when creating a window family, Revit will give you two possible UniFormat groups of the possible categories that a window can exist in. Here is a list of what a window could belong to:

B-Shell
    B20—Exterior Enclosure
        B2010   Exterior Louvers, Screens, and Fencing
        B2020   Exterior Windows or Screened Openings
    B30—Roofing
        B3020   Glazed Roof Openings and Roof Hatches

C Interior
    C1010 700   Interior Windows

Level 1 Major Group Elements

The system breaks building components down into seven distinct, alphabetically labeled categories that roughly resemble the sequence of construction, starting with (A) substructure, (B) structure and shell for building closure, then (C) interior construction, and (D) services; (E) equipment and furnishings follows thereafter. The last two categories are less sequential: (F) special construction, demolition, and (G) site work.

UniFormat categories

A. Substructure—foundation and basement construction
B. Shell—superstructure and building enclosure
C. Interiors—interior construction, stair, and interior finishes
D. Services—all mechanical, electrical, plumbing, fire protection, and conveying systems
E. Equipment and Furnishings—all equipment and furnishings
F. Special Construction and Demolition—construction not covered by other categories and demolition
G. Building Site Work—any work relating to site work, including services

Level 2 Group Elements

Level 2 is more specific, often relating directly to Revit categories that also deal with components associated with specific trades or contracts.

A. Substructure

This section specifies work done in the ground to create a structural foundation to carry the building above. It includes the slab-on-grade and any basement, as these are treated differently from other structures due to contact with the soil. This section includes site work such as the excavation and site prep work necessary for a basement.

A10   Foundations—all piles, footings, beams, slabs and walls, and other related foundations, including drains and insulation

A20   Basement Construction—excavation, shoring, backfill, and walls, including insulation and interior skin

B. Shell

This section includes all structure above ground as well as the exterior walls and roofs that close the building from the elements.

B10   Superstructure

B20   Exterior Enclosure

B30   Roofing

C. Interiors

This section includes all permanent static interior components specified by the architect or interior designer.

C10   Interior Construction

C20   Stairs

C30   Interior Finishes

D. Services

This section includes all components that help the building operate, including conveying systems such as elevators.

D10   Conveying

D20   Plumbing

D30   HVAC

D40   Fire Protection

D50   Electrical

E. Equipment and Furnishings

This section includes all the equipment and furnishings that go into the building.

E10   Equipment
E20   Furnishings

F. Special Construction and Demolition

The special construction includes all non-traditional building types that have special requirements. It also includes demolition and removal of hazardous materials.

F10   Special Construction
F20   Selective Building Demolition

G. Building Site Work

This section deals with all the site work like clearing, excavation, grading, and hazardous waste remediation. It includes roads and paving, exterior signage, bridges and tunnels, parking lots, planting and furnishings, and other landscaping. It also includes all services located on the site such as water services, heating and cooling, and electrical utilities.

G10   Site Preparation
G20   Site Improvement
G30   Site Mechanical Utilities
G40   Site Electrical Utilities
G90   Other Site Construction

MASTERFORMAT

MasterFormat is a specification system developed by the Construction Standards Institute. The first version containing 16 divisions was published in 1963.[2] Whereas UniFormat focuses on assemblies, MasterFormat is work results driven and tends to be broken down by materials rather than assemblies. The design team uses it to create the specification book that will function as a manual for the project. In 2004, MasterFormat underwent an overhaul and a new version was released that includes a total of 50 divisions; however, some are left as undefined for future use. The latest update was published in 2014. For more information, see: www.csinet.org/masterformat

Most of the materials in the VDC Revit template are classified using the MasterFormat system. See the Revit template on materials earlier in this chapter.

Divisions

00   Procurement and Contracting Requirements

01   General Requirements

Construction Subgroup

02   Existing Conditions

03   Concrete
- Forming and Accessories
- Reinforcing
- Cast-in-Place
- Precast
- Cast Deck and Underlayment
- Grouting
- Mass
- Cutting and Boring

04   Masonry
- Unit
- PStone Assemblies
- Refractor
- Corrosion-Resistant
- Manufactured

05   Metals
- Structural Framing
- Joists
- Decking
- Cold-Formed Framing
- Fabrications
- Decorative

06   Wood, Plastics, and Composites
- Rough Carpentry
- Finish Carpentry
- Architectural
- Structural Plastics
- Plastic Fabrications
- Structural Composites
- Composite Fabrications

07   Thermal and Moisture Protection
- Damp and Waterproofing
- Thermal Protection
- Roofing and Siding Panels
- Membrane Roofing

- Flashing and Sheet Metal
- Roof and Wall
- Specialties and Accessories
- Fire and Smoke Protection
- Joint Protection

08 Openings

- Doors and Frames
- Specialty Doors and Frames
- Entrances
- Storefronts and Curtain Walls
- Windows, Roof Windows
- Hardware
- Glazing
- Louvers and Vents

09 Finishes

- Plaster and Gypsum Board
- Tiling
- Ceilings
- Flooring
- Wall Finishes
- Acoustic Treatment
- Painting and Coating

10 Specialties

- Information
- Interior
- Fireplaces and Stoves
- Safety
- Storage
- Exterior
- Other

11 Equipment

- Vehicle and Pedestrian
- Commercial
- Residential
- Food Service
- Educational and Scientific
- Entertainment and Recreation
- Healthcare
- Facilities Maintenance and Operation
- Other

12 Furnishings
- Art
- Window Treatments
- Casework
- Furnishings and Accessories
- Furniture
- Multiple Seating
- Other

13 Special Construction
- Special Facility Components
- Special Purpose Rooms
- Special Structures
- Integrated Construction

Special Instrumentation

14 Conveying Equipment
- Dumbwaiters
- Elevators
- Escalators and Moving Walks
- Lifts
- Turntables
- Scaffolding
- Other

Services

21 Fire Suppression
22 Plumbing
23 Heating Ventilating and Air Conditioning
25 Integrated Automation
26 Electrical
27 Communications
28 Electronic Safety and Security

Site and Infrastructure

31 Earthwork
32 Exterior Improvements
33 Utilities
34 Transportation

Process Equipment Subgroup

40 Process Integration
41 Material Processing and Handling Equipment
42 Process Heating, Cooling, and Drying Equipment

43   Process Gas and Liquid Handling, Purification and Storage Equipment

44   Pollution and Waste Control Equipment

45   Industry-Specific Manufacturing Equipment

46   Water and Wastewater Equipment

48   Electrical Power Generation

OMNICLASS

The OmniClass Construction Classification system attempts to combine UniFormat, MasterFormat, and another format called EPIC (Electronic Product Information Cooperation).

It is being incorporated into the National BIM standards, as it is meant to form a classification structure for databases. The intent is to classify the built environment and the associated processes. OmniClass is currently in the development phase, and only certain tables have moved beyond draft to become a National Standard. As OmniClass is not a complete system at the time of writing this book, it will not be elaborated further than the different tables it includes. For more information regarding OmniClass, see: www.omniclass.org

The OmniClass tables are as follows:

Table

11   Construction Entities by function

12   Construction Entities by form

13   Spaces by function

14   Spaces by form

21   Elements

22   Work Results

23   Products

31   Phases

32   Services

33   Disciplines

34   Organizational Roles

35   Tools

36   Information

41   Materials

49   Properties

NOTES

1   "Background on Uniformat II: The ASTM E1557 Building Standard." UNIFORMAT II. UNIFORMAT, n.d. Web.

2   Warren Clendining, "History of Specifications." MasterFormat 2004 Overview— Technical Expressions Inc., July 13, 2009. Web. November 22, 2014.

5.5 VDC Standards

The process of design and construction very much evolves from the generic to the specific. Initially, a project is designed schematically-with a broad layout and generic decisions regarding specific building systems. More specific choices are made as the project develops, resulting in a detailed design intent specification outlining the project to be built. A design team's deliverable is 100 percent construction documents. For most projects, this deliverable forms the framework upon which the contractors base their bids. The awarded contractor provides these construction documents to specific trade contractors, who produce minutely detailed shop drawings of the specific components to be installed.

The VDC process must follow that very same concept. During the design process, modeling anything beyond a specific level of detail too soon in the process is useless and even risky. Not only does over-modeling indicate specifics that have not yet been decided, it also tends to take more time to implement changes once details are determined later on. An uneven level of detail also creates issues when using the model for the different types of VDC services. In order to maintain good-quality control and address the order of model creation, the American Institute of Architects developed a system called "Level of Development" (LOD) in 2008. The AIA also assembled a document called E202-2008 to tackle not only when a component should be modeled, but by whom.

This document proved to be a good start, but major gaps arose when running a true VDC project from design to construction and ultimately operations. In 2011, BIMForum launched an initiative to evolve the system to address these issues. This resulted in the 2013 LOD specification, which is built on E202, but goes into greater detail regarding to what level each component should be modeled. The main difference is that it clearly illustrates what and when sub-components should be used, clarifying each team's expectations when using a received model. LOD specification is also essential for other VDC services that utilize the model such as cost estimating, 4-D scheduling, clash detection, and constructability review.

LEVEL OF DEVELOPMENT 2013

The LOD system spans the three main phases of a building project: design, construction, and operations. A numbered system is used to identify each phase. Design intent models are 100 to 300, construction models are 400, and operations are 500. LOD 2013 introduced LOD 350, a further

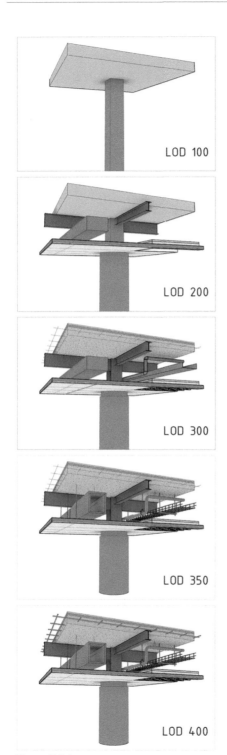

Figure 5.5.1
LOD illustrations for each LOD stage

developed design model. LOD 350 grew out of the realization that the previous design model was not sufficiently detailed to run effective clash detection, which is necessary in order to properly coordinate the ducts and pipes as hangers, and other detailed parts introduced later during the construction phase. This omission resulted in new clashes during the construction phase, necessitating change orders and calling into question the purpose of running clash detection during the design phase. LOD 350 uses the UniFormat system to classify an object and list each one with model illustrations of typical detail levels for each level of development. For additional information, see the BIMForum website: https://bimforum.org/lod

Design

•	Schematic	100
•	Design Development	200
•	Construction Documents	300
•	Coordination Design Model	350

Construction

•	Construction Model	400

Operations

•	Facilities/Operations Model	500

LOD 500

The LOD 500 model is sometimes referred to as the as-built model or facilities management model. Currently, there is no clear consensus of what exactly is included in that model, so LOD 500 was not included in LOD 2013. Broadly defined, it is a model that reflects what was actually built. The model should include all elements pertinent to the facilities manager, including information regarding manufacture, manuals, and data relevant for the end user. All rooms and levels should be correctly defined in the model, as taking advantage of Revit's locational database to automatically track and locate different components is one of the greatest benefits to embedding data in the model for facilities management. It is also useful to connect an external database to the model to make data entry more efficient. The Construction and Operations Building Information Exchange (COBie) is becoming the standard format for transferring VDC data into facilities management software.

IFC INDUSTRY FOUNDATION CLASSES

Industry Foundation Classes (IFC) is an open, non-proprietary geometry, and information exchange standard being developed to ease project collaboration for design, construction, and operations models across different software platforms. Most VDC software supports this standard. The latest version, as of this writing, is version 4 released in 2013. IFC is being developed by the BuildingSMART alliance. The core concept of their effort is expressed by their logo's four intertwined squares which stand for design, procure, assemble, and operate. Previously, it was only possible to import IFC models into Revit, but starting with the release of Revit 2015, it is possible to link IFC files to a Revit project, enabling seamless integration of updates to the IFC model.

COBIE

Construction-Operation Building Information Exchange (COBie): The COBie standard is an attempt to create a standard for how facilities management data get incorporated into information models so that design and construction teams could continuously input the relevant information as a project proceeds. Successful usage of the COBie standard requires the end user to specify the necessary data, as the COBie standard is so broad in definition.

More facilities management software applications are releasing VDC integration products. Autodesk has a free plugin for data export to the COBie standard that works quite well.[1]

For more information regarding COBie, see: www.wbdg.org/resources/cobie.php

NOTE

1 Rich Mitrenga and T. J. Meehan, "COBie Toolkit for Autodesk Revit." YouTube, May 19, 2014. Web. November 23, 2014.

6 The Future of VDC

We look at the present through a rearview mirror; we
walk backwards into the future.

(Marshall McLuhan)

In the coming years, advancements in a wide range of technologies will
have a large impact on the phases of building: conception, design,
construction, and maintenance. Traditionally, buildings are a manifestation
of physical drawings and models. In the future, performance-based design
will enable myriad scenarios to be simulated, tested, and validated in
advance of construction. Consequently, designs will be better informed,
and fewer errors will present themselves in the field. Real-time systems
that track the state and location of materials, possess the ability to
dynamically control automated construction robots, and interface with
various sensors will substantially optimize construction.

This chapter will focus on tools that help to visualize, automate, and
analyze a building, not just as a black box, but taking into account all the
individual parts and assemblies that comprise it. Some of the technologies
and concepts mentioned in this chapter already exist in one form or
another, either within the AEC industry or within other industries.
However, it is when these tools are integrated with existing AEC workflows
that the full potential of VDC will be unlocked. That AEC is one of the few
domains yet to be disrupted by the digital revolution is symptomatic of a
culture that is slow to adopt new hardware, software, and workflows.
Adoption of VDC will become increasingly critical with increased quality
and decreased cost. Owners and developers, being the main beneficiaries,
will be the parties most responsible for instigating this change.

With advanced visualization tools and data, such as augmented reality,
virtual mockups, and complete building simulations, designs can be
virtually realized and optimized for multiple factors such as usability,
energy consumption, environmental, aesthetics, constructability, material
availability, operational, and lifecycle considerations.

Developments in automation and fabrication will allow for faster and
more precise production. Some technologies might eliminate the need for
humans to perform tasks that are repetitive or monotonous, while others,
such as exoskeletons, will serve to extend a site worker's capabilities.

Figure 6.1
Diagram of the future of VDC. The software, hardware, and workflow advancements, and the concepts that connect them

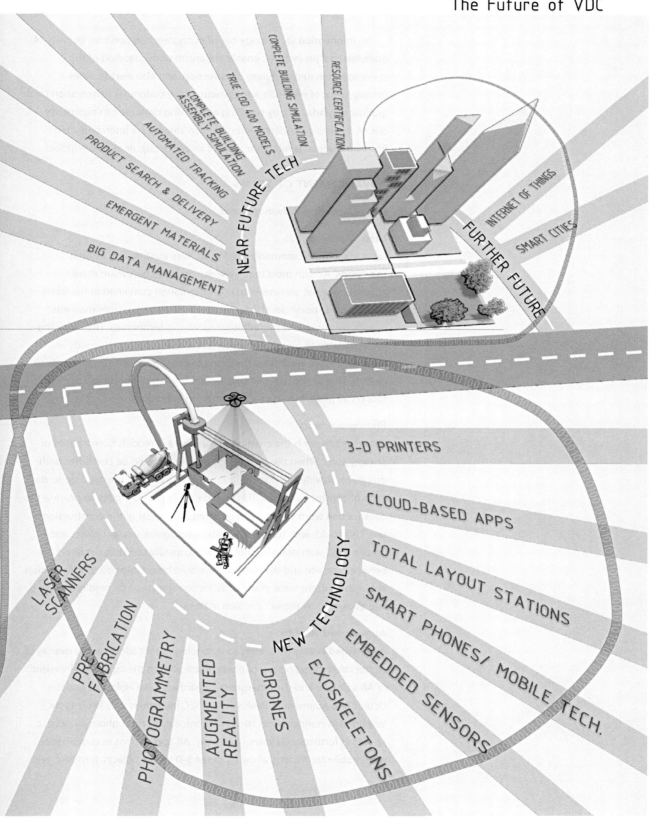

NEAR FUTURE TECH

COMPLETE BUILDING SIMULATION
RESOURCE CERTIFICATION
TRUE LOD 400 MODELS
COMPLETE BUILDING ASSEMBLY SIMULATION
AUTOMATED TRACKING
PRODUCT SEARCH & DELIVERY
EMERGENT MATERIALS
BIG DATA MANAGEMENT

FURTHER FUTURE
INTERNET OF THINGS
SMART CITIES

3-D PRINTERS
CLOUD-BASED APPS
TOTAL LAYOUT STATIONS
SMART PHONES/ MOBILE TECH.
EMBEDDED SENSORS

NEW TECHNOLOGY

EXOSKELETONS
DRONES
AUGMENTED REALITY
PHOTOGRAMMETRY
PRE-FABRICATION
LASER SCANNERS

An information technology-based ecosystem can optimize delivery of materials to a project site; enable enhanced understanding of the condition of materials during construction; and also alert facilities management of materials and systems as the building is in operation (and gradual degradation). By synching and linking data, we can potentially obtain a cohesive understanding of how "things" are interconnected—from a molecular level to the scale of an entire city, or even the globe.

ADVANCEMENT OF TOOLS

Field-Capturing Advancements

Laser Scanning

Laser scanning is commonly used to capture existing and as-built conditions. As with most hardware, scanners are becoming more affordable, faster, and easier to use. Information contained in the laser-captured, 3-D point cloud is far more complete and reliable than was previously possible with traditional surveying techniques. The modeling tools available to generate information models from point clouds are becoming more automated, reducing the cost to create accurate as-built models. Due to the reduced cost, laser scanning will become a standard tool on all projects that deal with existing conditions.

Photogrammetry

Photogrammetry is the creation of maps or 3-D models from a series of photographs. When the technology matures it could be combined with drones to provide the construction control team with the most up-to-date state of construction sites. This would enable the possibility to provide the construction team with a daily, updated 3-D model of the construction site. This could, with further innovation, recognize similar objects and update itself with exact location data and installation dates. A direct link between the site and the main office would be established, as information about the current state of the construction site could be readily shared and evaluated by project decision makers.

Augmented Reality

Augmented Reality (AR) refers to technologies that allow for the overlay of digital information on top of physical space. With the rapid development of AR software and the emergence of hardware like HoloLens and the Oculus Rift, Augmented Reality in the AEC industry is very likely to be ingrained in standard operations and procedures throughout all project phases in forthcoming years.[1] Currently, AR applications in combination with mobile technology allow us to see 3-D virtual objects, furniture, and

Figures 6.2A–C

"Flight Assembled Architecture" is the first architectural installation assembled entirely by flying robots. "Flight Assembled Architecture" consists of over 1,500 modules, which are placed by a multitude of quadrotor helicopters, collaborating according to mathematical algorithms that translate digital design data to the behavior of the flying machines

Credits: Gramazio & Kohler and Raffaello D'Andrea in cooperation with ETH Zurich. Client: FRAC Centre (Co-Producer). Collaborators: Andrea Kondziela (project lead), Sarah Bridges, Tim Burton, Thomas Cadalbert, Dr. Ralph Bärtschi, Peter Heckeroth, Marion Ott, Tanja Pereira, Dominik Weber, Dr. Jan Willmann. Selected experts: Wind Tunnel Testing: Chair of Building Physics, Prof. Jan Carmeliet, ETH Zurich and Empa. Structural Engineering: Dr. Lüchinger + Meyer Bauingenieure AG. Façade Engineering: Dr. Lüchinger + Meyer Bauingenieure AG. Energy Consulting: Amstein + Walthert AG. Sponsors: Pro Helvetia Swiss Arts Council, Centre Culturel Suisse Paris, Platform, Regroupement des Fonds régionaux d'art contemporain, Vicon Motion Systems, ERCO Leuchten GmbH, JET Schaumstoff-Formteile GmbH

fittings visualized in physical space, as well as virtual information models overlaid on printed 2-D plans. In the future, AR in conjunction with location sensors and GPS will make it possible to superimpose planned construction onto existing conditions through an AR viewing platform. Pre-programmed animations could show team members sequences of installation, and guide them through complex assemblies.

Drones, or Unmanned Aerial Vehicles (UAVs)

Equipped with cameras and GPS, drones can be programmed to follow predefined routes, periodically capturing the state of construction sites and perform inspection of places that are hard to reach. Combined with software developments such as photogrammetry (the science of making measurements through photographs), drones can augment current inspection processes and automate tracking certain tasks. Additionally, improved precision, strength, and reliability will allow drones to perform physical tasks, including moving materials and placing objects in defined locations.

3-D Printing Advancements[2]

3-D printing has generated a lot of attention in recent years. Small-scale printers are becoming more affordable and enable design and construction teams to prototype ideas. Future developments anticipated in 3-D printing include an increase in the printer scale and in the diversity of materials that can be printed. Current experiments with the technology include 3-D printed structural steel joints, earthquake-resistant columns, and entire concrete structures. Combined with advances in robotics, 3-D printing has the potential to revolutionize how buildings are designed and constructed. The technology also opens up the possibility of mass customization, eliminating design constraints currently imposed by the need to use standard components.

Exoskeletons

Robotic exoskeletons assist builders in maneuvering heavy objects so that they can focus on putting objects in place precisely. These suits are being tested for ship manufacturing in Korea.[3] On a construction site, an exoskeleton could potentially be used to guide a worker to the correct

Figures 6.3A–B
(facing page) 3-D printed building modules can create seismically resistant structures. Interlocking components diffuse the force of an earthquake, a concept drawn from traditional Incan ashlar techniques. The "Quake Column" is a concept designed by Ronald Rael, Virginia San Fratello. Material: Sand. Dimensions: 6'-5" tall

installation point, help with heavy lifting, and provide a range of built-in tools. A simulation of the building assembly in the information model could include the path a worker must follow, which could be programmed into the robotic suit. Exoskeletons are currently extremely expensive, but, as with all new technologies, the price will drop as the technology is developed.

Integrated/Embedded Sensors

Embedded sensors can expose the state of building materials. For example, the curing of concrete can be monitored through embedded sensors for better quality control. As buildings age, materials tend to deteriorate and become structurally inefficient. Embedding sensors in concrete enables engineers to monitor the condition of structural members in a building, alerting them to necessary maintenance issues. Sensor networks, which can monitor the built and natural environment, are integrated with communication technologies, signal processing, and statistical models for network data analysis, enabling the development of more sustainable systems.

BUILDING INDUSTRY ADVANCEMENTS

Automated 3-D Layout Total Stations

Traditionally, the layout of building elements is a manual translation from 2-D drawings to the site. There are frequently errors and inconsistencies in this process. As the construction information model is a virtual copy of the completed building, with some preparation, robotic total stations can be connected to the information model to guide the physical layout on the jobsite. 3-D total stations require only one person to operate, they minimize mistakes and increase accuracy, resulting in a faster layout while also verifying the installed building elements. As the technology is directly connected to the information model, total stations will become a tool for verifying as-built conditions on a day-to-day basis, so the as-built model can be built as construction progresses rather than after the construction is completed, which is often the case today. In the future, versions of this machine could potentially be used for continuous tracking purposes, ensuring construction materials are placed in the right locations in the right quantities, and on time.

Complete Building Simulation

The ability to visualize and simulate the entire lifecycle of buildings, from construction through operations, and finally demolition, will alter how buildings are designed, how they will be occupied, and how they perform.

For example, if all building materials in the future will be recycled, how will buildings be assembled to facilitate demolition and recycling? How will they look and feel? Certainly more and more design problems will be solved by computation. A range of systems can already be simulated, including lights, people, heat, material properties, energy usage, and wear and tear. As computing power increases, multi-objective optimization will be an integral part of the design and execution processes. Buildings will become less about a series of compromises as to what is possible, and more about fully, precisely calculated solutions.

True LOD 400 Models

True LOD 400 models include every screw, nut, and bolt. Currently, producing an LOD 400 model is a labor-intensive, time-consuming process. Rarely is an information model developed to this level of detail, except in highly specialized fabrication modeling. However, with advances in modeling software, hardware capabilities, contractors' increased familiarity with information modeling, and 3-D scanning of existing conditions, it will become more feasible to develop a true LOD 400 model for every project.

Complete Building Assembly Simulation

Complete building assembly simulation refers to completely coordinated and planned assembly simulation, in which intelligent algorithms designed to analyze numerous scenarios can assist in the team's decision making and selection of the most efficient scheme of assembly. For example, most of the work of routing, analysis, and modeling of mechanical systems may be performed by a 3-D algorithm, which tests millions of routing options in order to find the optimum one, based on choices such as clearance, minimum material usage, energy performance, installation, or other criteria. Installation information from each manufacturer can also be included to ensure systems are installed in the correct sequence.

Prefabrication Advances

With a perfectly coordinated model and computerized robotic manufacturing methods, it is possible to increase the level of prefabrication on a project. Prefabrication increases accuracy, as it happens in a controlled environment not dependent on climate and other processes that can disrupt manufacturing on-site. It demands planning of the process of delivery and installation in detail. There is a trend among subcontractor trades to take on more than one trade, as they can build multiple trades as assemblies and put them together on-site. In fact, whole

Figures 6.4A–B

Photos of the Hy-Fi organic pavilion at MoMA PS1 from June, 2014, designed by the firm The Living. The bricks are entirely organic, made largely from corn stalks, and mycelium, and are ultimately compostable

Photo credit: Christina Kissel

sections of systems can be prefabricated and simply installed on-site, transforming the site from a construction site into an assembly site. This will mitigate delays that are often caused by having too many trades working in the same place on a construction site.

Automated Tracking

Automation will continue to develop for tracking on-site work and materials, through the use of existing technologies, like RFID tags on individual building components, and with newer sensor technologies that will allow for more intelligent tracking. "Just in time" and mass-customized fabrication scenarios will become more prevalent, as information about what has been installed on-site becomes available in real time. Equipment and material movements can also be tracked on-site, paving the way for advances in self-assembly and robotic fabrication.

Automated Product Search and Delivery Planning

An information model can include specifications for each building element, and if this data is connected to the web, it can be connected directly to manufacturers. This will make it possible to search for the optimum product based on desired criteria such as performance, price, durability, delivery method, distance, quality requirements, installation method, lead time, or environmental impact. Advice on relating products that perform similarly will also be available. Each component will be tagged and tracked for just-in-time delivery.

Emergent Materials

The development of new materials is rapidly evolving. The industrialization of materials such as graphene, which is 100 times stronger than steel and an excellent conductor of electricity with almost no resistance, opens up new possibilities. Graphene is but one example of many new materials being developed with unique properties. These materials will result in new ways of conceiving buildings, for example, by providing new ways to simultaneously insulate and ventilate buildings, or harvest energy from the environment. Other materials, such as carbon, can be programmed to bend or twist, and modules can be made to self-assemble.[4]

Organic examples of emergent materials include mycelium, a fungus that, when dried, can be used to form a super-strong, water, mold, and fire-resistant building material.[5] This material can be grown and formed into almost any shape, and has inspired prototypes of "growing buildings," by growing the fungus in formwork. How to calculate and analyze a building that is "alive" or "growing" will be a challenge to VDC processes and workflow.

Resource Certification

Connecting databases of materiality information makes it possible to evaluate the environmental impact of material selections from the perspective of the entire building lifecycle. These types of analyses will become essential to the next generation of sustainability certifications. The complete supply chain will be taken into account in the sustainability analysis. Environmental considerations would include end of life and possible reuse. For example, a Revit plugin connects building components to a material database, so short feedback loops can inform the designer about the environmental impact of different material selections.[6]

GENERAL TECHNOLOGICAL ADVANCEMENTS

Mobile Technology

Advancements in mobile technology align with the needs of construction professionals with the need to access and capture increasingly large amounts of data while on-site; mobile devices are evolving rapidly in computing power, storage capacity as well as available applications. Mobile devices are equipped with successively high-resolution screens and cameras, and already, certain cell phones include 3-D scanners.[7] The ability to access all project information from a small, powerful, and yet affordable device has huge implications for project delivery in the field.

Internet of Things

Another application of sensors is referred to as the "Internet of Things." The emergence of tiny, inexpensive, energy-efficient sensors with wireless transmitters makes it possible to track almost any part of the environment. Embedding wireless sensors into objects and connecting them to the internet creates an internet of smart objects. Tracked items could be anything from a carton of milk in the refrigerator (so that it can be known when milk is about to expire) to the condition of concrete (to indicate its state during the curing process). Home automation is making its way into the consumer market. Computer giants like Google and Apple are in the process of developing standards that will enable them to build automation for systems such as heating and cooling.[8]

The functionality of such systems should be included in the model in order to optimize their placement and function.

Cloud Computing

Cloud computing allows a greater amount of data to be saved, synced, and accessed instantly across multiple devices. As internet connection is becoming ever faster and more reliable it is possible to move software, data, and computational tasks from in-office computers and servers to servers in the cloud, making data accessible from multiple locations and enabling access to multiple types of devices. Centralized processing power enables cloud-based solutions to democratize technology, making it more accessible to the project team by alleviating the cost of equipment and specialized software, and increasing the project team's ability to collaborate from any location. Software upgrades can be done gradually, which eliminates the need for an IT department to schedule maintenance. The cloud offers almost infinite processing power which is ideal for large, complex calculations. Elimination of paper documentation in lieu of cloud-stored and cloud-processed data is transforming construction workflows.

Big Data

Recent innovations in software, hardware, and tools have focused on enhancing productivity, construction workflows' primary objective. A unified database format for project data would enable the industry to collect and store information from various project topologies. When enough data are collected and analyzed, decisions that are now experience-based will instead be informed by the cumulative analysis of other projects—"big data" for construction.[9] The idea that these data can be a real resource and asset for the building industry is becoming more commonplace, especially on larger projects with many phases, items, and materials to track. Additionally, sectors of the industry that utilize building data long after project construction is complete are finding information model data and ongoing mechanical and systems data very useful for facilities management purposes.

Smart Cities

More than half of the world's citizens now live in a city. According to the United Nation's World Population Index, approximately 66 percent of the growing population will live in urban environments by 2050.[10] This means that about 2.5 billion more people would call a city their home.

A modern city is an extremely complex organization that requires a well-planned infrastructure in order to function properly. Smart Cities is the term for the integration of information technology with the urban fabric. Innovations in civic and transportation data standards will enable the development of software that will allow for greater control of the systems that drive the city. New building forms will not just consume resources like electricity, water, and food; some will actually produce resources to feed back into the urban system. Trends such as urban farming and micropower could accelerate the city's transformation into a neural network, organized by nodes of interdependent systems. The systems of the city are becoming technologically integrated, adaptable, and agile.

There are many current experiments with Smart Cities and city data. Berlin, Montreal, and Kyoto all are creating 3-D city models.[11] Rio de Janeiro has created a central control room as a hub of all of the data collected throughout the city, integrating data from various sources and agencies.[12] New cities that are built from the ground up will integrate VDC technologies as a basis of civic operations.

CONCLUSION

Construction is one of the oldest professions whose fundamental nature will not change as a result of technological innovation; projects still need to be designed, planned, and constructed. However, the approach and tools for those processes will be different. Traditional tools are fragmented and not sufficient in describing the complex buildings of today and tomorrow. The communicative and interconnected nature of VDC will enable project participants to have a much higher level of control. Successful implementation of VDC is hinging on how well developing technologies interact with design and construction needs.

The examples mentioned in this chapter highlight some of the trends that will drive this shift. The biggest theme leading this change is centered on data and analysis. Buildings should not be solely defined by the sum of their parts, but by the parts and assemblies themselves and their roles within greater systems and processes—from conception, 3-D creation, through to construction and facilities management. The "building blocks" of the future are represented by the data they embody, and not merely by their physical presence. The boundaries between the virtual and real are increasingly blurred as technology becomes part of the very fabric of design and construction.

It is an exhilarating time to be involved in the design and construction of buildings and cities.

NOTES

1   David Barista, "Augmented Reality: 12 Applications for Design and Construction Professionals." *Building Design + Construction*, September 4, 2014. Web. November 2, 2014.

2   Ian Harvey, "Printing Structures the next Step in Construction Technology?" *Daily Commercial News*, June 18, 2014. Web. October 1, 2014.

3   Shipping company Daewoo has fabricated RoboShipbuilder, an exosuit being used by employees that can lift up to 30 kg (66.1 lb). Mick, Jason. "Korean Shipbuilder Uses 'Iron Man' Exosuit to Help Build World's Largest Freighter." *DailyTech*, August 4, 2014. Web. October 1, 2014.

4   Modular units created by MIT's Self-Assembly Lab have specially configured geometry and attraction mechanisms that ensured the units came into contact with one another and auto-aligned into locally correct configurations. "Self-Assembly Lab." Self-Assembly Lab, n.p., n.d. Web. November 5, 2014.

5   Mark Boyer, "Philip Ross Molds Fast-Growing Fungi Into Mushroom Building Bricks That Are Stronger than Concrete." *Inhabitat*, June 25, 2014. Web. November 22, 2014.

6   "Tally(r) Revit Application." *Tally(r) Revit Application*. n.p., n.d. Web. November 5, 2014.

7   In 2014, Google announced its Project Tango, which equips mobile devices with customized software and hardware to track the 3-D world around them. Sensors in a Project Tango device can make over a quarter of a million 3-D measurements every second, updating its position and orientation in real time: www.google.com/atap/projecttango/#project

8   Ashish Gulati, "Long-term Evolution and the Internet of Things." *Dataquest*, August 20, 2014. Web. September 24, 2014.

9   Andera Al-Saudi, "The Future of BIM Must Focus on Data Not Productivity." BIM Hub, n.p., August 10, 2014. Web. November 5, 2014.

10 "World's Population Increasingly Urban with More than Half Living in Urban Areas." UN News Center. United Nations, July 10, 2014. Web. November 2, 2014.

11 "3-D City Models Expose Cities to More Growth—3-D Visualization World, " n.p., June 8, 2012. Web. November 5, 2014.

12 Alex Marshall, "Data: Are Cities Losing Control Over 'Smart' Initiatives?" (opinion). *Government Technology*, March 4, 2014. Web. November 5, 2014.

Glossary of VDC Terms

4-D Schedule A visualization of the sequencing of construction created by linking schedule activities to corresponding model components.

3-D Printing A process which generates 3-D objects from a virtual model.

AEC Architecture, Engineering, and Construction.

Analysis Software Software that is designed to perform various engineering calculations and simulations. Typical analysis software contains optimization routines that are able to iterate through many scenarios enabling qualitative engineering analysis.

Augmented Reality A process which overlays computer-generated graphics or information over the physical reality, allowing a user to contextualize computer-generated information in the real world.

Authoring Software Software that is used to create the information model which includes the generation of the project's geometry and base information attached to geometrical elements.

BIM Building Information Modeling.

CAD Computer Aided Design.

Central File Resides on a server with which local user files communicate. It keeps track of changes as well as user rights.

Cloud Computing Applications and data are stored on remote servers "in the cloud" rather than locally on the user's hardware. This enables concurrent workflows and is ideal for rapidly evolving software.

CM Construction Manager.

COBie Construction-Operation Information Exchange A data standard which handles porting data from the design and construction phase into facilities management software.

Construction Model Fully coordinated composite model developed to the highest level of detail LOD 400. It includes models of each building system accurate in size, shape, location, quantity, and orientation. To be modeled by system by level.

Coordination Model Information model that is created for the purpose of resolving interferences between various systems present in a project. Central model which holds the structural, architectural, and systems sub-models.

CPM Schedule   Critical Path Method is a methodology of establishing what activities are critical to complete in order to start the next. The critical activities have to be done on time in order to avoid delaying the overall project.

CSI   Construction Specifications Institute.

Design Intent Model   Information model created to the level of detail contained in the construction drawing set. LOD 300-level model which includes architecture and structure as depicted in the construction documents.

Drawing Set   Collection of drawings compiled for the purpose of describing a project.

Drone   An unmanned aerial vehicle containing sensors or cameras. In the context of the construction industry it is often used to examine the state of the worksite, and inspect completed activities.

File Formats

DWF   Design Web Format—2-D and 3-D viewing formats that include BIM data.

DWG   AutoCAD drawing format—does not include BIM data.

NWC   Navisworks cache files.

NWD   Navisworks document format.

NWF   Navisworks file format—Native Navisworks format.

RCS   Recap format—Autodesk's point cloud format. Has become the preferred format for a Revit-centric workflow.

RFA   Revit Family (component) format.

RVT   Revit Project format.

SP   Synchro 4-D scheduling file. Includes the schedule and linked geometry.

IFC   Industry Foundation Classes—a software agonistic standard and BIM file format.

Information Model   A database-centric 3-D model created to high level of detail, which contains various types of data gathered during a process of project realization.

ISO   International Organization for Standardization.

Laser Scanning   A way of scanning 3-D space using a scanner that throws millions of laser beams in a spherical motion where the distance of each object the laser hits is measured. This results in a "cloud" of point which is then saved in a file, often called a "point-cloud."

LCA   Lifecycle Assessment.

Local Files   The file in Revit that the individual user works with which are synched with the central file so that changes can be shared among multiple users.

LOD   Level of Development (AIA-document E202 and LOD 2014), definition of what should be modeled during the different phases of a project and to what detail level.

Master Format   A system developed by the National Specification Institute which is primarily used for specifications.

MB   Megabyte—one billion bytes of information, used to quantify file sizes.

MEP   Mechanical, Electrical, and Plumbing (usually includes Fire Protection).

NBIMS   National BIM Standard. An organization and written guide that provides consensus-based standards for delivering best BIM practices for the entire built environment: www.nationalbimstandard.org

OmniClass   A classification system developed to be effectively used for databases which is being adopted by the National BIM Standard.

Photogrammetry   The science of making measurements from still photographs, particularly the exact location of surface points. Often used as a term for generating 3-D models from a series of photographs.

PLA   Project Labor Agreement.

Point Cloud   A file that contains scanning data collected from a laser scanner consisting of millions or possibly billions of points.

RFP   Request for Proposals.

TB   Terabyte—one trillion bytes of information, used to quantify file sizes.

Uniformat   A classification system developed by the Construction Specifications Institute. It divides the project into separate assemblies. Uniformat II is the most used version for cost-estimating purposes.

VDC   Virtual Design and Construction—the implementation of data-driven technologies and processes, including Building Information Modeling.

VDC Ecosystem   This defines all the interdependent systems of mechanisms that form a cohesive functional AEC VDC system.

VDC Process (or Methodology) VDC processes are workflows that incorporate the information model and integrate previously disconnected aspects of design and construction. VDC processes seek to apply new technologies to the AEC industry and link all the work being done by the project team into the information model.

VDC Product A VDC product is the deliverable resulting from a VDC service, such as a point cloud, a systems coordination model, database, or a constructability logistics animation.

VDC Service VDC services are specific services unique to VDC, such as clash detection, 3-D scanning, tracking, or information model authoring.

Virtual Reality An immersive experience where the virtual environment is experienced artificially.

Virtual Request for Definition (vRFI) A request for information that derives from issues found in the model. vRFIs are different to regular RFIs as they often refer to pure modeling issues.

Work Breakdown Structure Hierarchical organization of relationships in a project such as phases, deliverables, and work packages.

Index